Bags and Saddles

OTHER SAGEBRUSH LARGE PRINT WESTERNS BY

LAURAN PAINE

Buckskin Buccaneer
Guns of the Law
Six-Gun Atonement
The Californios
The Past Won't End
The Rawhiders
Valor in the Land

Bags and Saddles

LAURAN PAINE

Library of Congress Cataloging in Publication Data

Paine, Lauran.
Bags and saddles / Lauran Paine.
p. cm.
ISBN 1-57490-198-2 (alk. paper)
1. Large type books. I. Title.
[PS3566.A34B27 1999]
813'.54—dc21 99-30783
CIP

Cataloguing in Publication Data is available from
the British Library and the National Library of Australia.

Published by arrangement with Golden West Literary Agency.

Sagebrush Large Print Westerns are published in the United States and Canada by Thomas T. Beeler, Publisher, Box 659, Hampton Falls, New Hampshire 03844-0659. ISBN 1-57490-199-0

Published in the United Kingdom, Eire, and the Republic of South Africa by Isis Publishing Ltd, 7 Centremead, Osney Mead, Oxford OX2 0ES England. ISBN 0-7531-6015-3

Published in Australia and New Zealand by Australian Large Print Audio & Video Pty Ltd, 17 Mohr Street, Tullamarine, Victoria, 3043, Australia. ISBN 1-86442-248-3

Manufactured in the United States of America by Sheridan Books in Chelsea, Michigan.

Bags and Saddles

CHAPTER 1

SHERIFF TOM MCGRATH CONSIDERED THE EVIDENCE—the open gate, the tracks leading eastward, the boot-sole imprints, and finally, the pole where the missing saddle, blanket and bridle had been.

There wasn't need for much more evidence; as a matter of legal fact, Tom McGrath didn't need as much evidence as he had. The one prerequisite to the filing of charges of horse theft was that the horse be gone. It was. The next prerequisite was to find someone besides the complaining owner riding that horse.

Sheriff McGrath took his posse eastward out of the Meuller ranch yard following the tracks, determined to locate and apprehend the person riding the horse. He had four men with him, recruited from cowboys who'd been in town over at Puma Station when he'd heard there had been a horse theft. They were good men on a trail or good men in a tussle. Tom McGrath had been sheriff in Dunphy County eleven years; he knew every worth-while range rider for miles around. These four were top hands. Unless the horse thief sprouted wings overnight and flew away with Old Otto Meuller's horse, he would shortly be overtaken and hauled back to Puma Station.

Which was, as Tom McGrath told the four men with him, a darned sight better fate than might have overtaken him if a citizen posse had heard of his theft and had taken after him. Citizen posses had a habit of trying a man right where he was apprehended, then efficiently hanging him from the nearest tree limb. There was a lot of sentiment through the cow country in favor of these summary executions. So much favor, in fact, that there was hardly a cow town which didn't have men walking the roadways who had participated in such lynchings, and against whom no one would inform.

This time it would be different. The cowboys riding with Tom McGrath respected him. Tom hadn't reached forty yet—in fact, he wasn't yet pushing thirty-five—but he'd proven himself game, fair, uncommonly capable with his gun, and upon rare occasions with his fists. He'd been a range man for most of his earlier life, before taking on as a deputy sheriff in Dunphy County. After the death of Will Jamieson he'd stepped into the sheriff's boots and had been wearing them ever since.

Tom McGrath was of medium height, heavy-shouldered and slab-hammed. He had dark curly hair and grey eyes. He had the endurance, the stamina and doggedness of a bull-buffalo, and although he smiled easily, when he *didn't* smile folks said it was a good time to shut up and just listen.

On this trail, as on other similar trails, Tom McGrath read sign like an Indian. For instance, after they'd passed around the northern end of Puma Station, still traveling east, Tom told his possemen there was something wrong with the horse thief; either he was sick or dog-tired, because he let the horse pick its own way, made no attempt to hit hardpan where tracks wouldn't show, and finally, because the horse thief wasn't riding fast.

One of the range men was unimpressed by that last factor, and threw Tom a droll little grin. "Why should he make tracks? After all, he done stole that horse last night, so he's got a long four or five-hour lead on us.

"Two hours," stated Tom, which made all the cowboys look at him. "If you fellers'd looked back at the Meuller place instead of ogling old Otto's two big daughters, you'd have seen the dew had already settled when he rode out of the barn. That would put him only a couple hours ahead, and—" McGrath pointed to the sign they were following—"he's not widening that gap one bit. Look there; the horse stopped to grab a mouthful of grass. No professional living lets a horse eat while he's riding it."

One of the rangemen jerked up his head. "Hey," he said quickly, "this could be a kid we're shagging. This could be some doggoned runaway kid."

The others turned this over in their minds as they went along, watching the tracks. Tom McGrath offered no opinion one way or another. For perhaps a mile, or slightly more, the possemen discussed this possibility. They asked Tom. He only shrugged and kept studying the land ahead.

If the horse thief was a kid, then he was a green one to the wide open spaces of Dunphy County, Montana Territory. Tom McGrath knew every range, every rocky slope, every mile of rolling prairie in his domain; if a horse thief also had known the country, he'd be heading north toward the Canadian line where he could shake any pursuers—not straight east across the low brakes above Puma Station, which ended ten miles farther along in pitilessly open prairie country, where a man with just one eye could see for a hundred miles all around.

Tom didn't heed much of the speculation his riders indulged in. He was beginning to build a composite notion of the horse thief up ahead: lean, hungry-looking, either sick or so tired he had to force himself along. Probably a young man; younger than Tom himself. Maybe a down-and-out range rider or perhaps a novice outlaw. But not a kid; definitely not a kid. Whether the others had noticed or not, they had ridden past two brown-paper cigarette butts.

They were well northeast of Puma Station, and the land was beginning to change subtly, when the tracks began to veer southward. One of the cowboys nodded at that. "He's got a good look at the prairie out there an' figured he could be seen crossing all that nothing', an' has now decided to cut off in a fresh direction where a feller might find some shelter."

Tom accepted that. They duly turned and pushed on. It was past noon now; the springtime day was balmy, with a hint of suppressed cold. If they didn't conclude the chase by nightfall, there were going to be some disgruntled possemen. They each had a rider's jacket lashed aft of the saddle cantle, and they wore Pendleton woolen shirts buttoned to the throat, but Montana in springtime after sunset was as cold as the bottom of a well.

"He ain't hurryin' even yet," observed a puzzled cowboy. "Tom, what say we give him a run to close the gap a mite?"

McGrath shook his head. "We'll get him directly without worrying the horses. He's going slower and slower. That horse has figured out its rider. It's just shuffling along."

The range man squinted down across the little broken hillocks and arroyos as he said, "I'm almost sorry for him. No one ought to be that green."

"Well," opined another tough-faced rider, "I'm sure you'll agree, he's a long shot from bein' smart or experienced at stealin' horses. The idiot's headin' straight down parallel to town. He can't have very good notions of direction, either."

McGrath had the answer to that. “He passed Puma Station long after midnight. There weren’t lights or noise or enough stars last night for him to see the town from up where he was. Sure he’s parallelin’ town now—but he doesn’t know it. I’ll tell you somethin’ else he’s doing: He’s keeping an eye peeled for game. He’s hungry.” McGrath pointed. “The tracks turn now and then; he’s looking for a rabbit or a sage hen to shoot.”

One of the men screwed up his face in concentration, then came up with something relevant. “Sure he’s hungry. Even if he ate only an hour or less before he stoled old Meuller’s horse, he’d still be hungry now because he’s been ridin’ the equivalent of a full day without grub.”

“How do you know that?” asked a skeptic.

The cowboy said: “Easy, you darned fool. He also stole a saddle and bridle an’ blanket from Meuller. He didn’t even have his own outfit, which means he didn’t have grub with him.”

There was a wrangle over this. The skeptic said their horse thief could have had grub in his pockets. The other man said then why hadn’t they seen any empty tins along the trail? It went on for nearly ten minutes, see-sawing back and forth. Not that either cowboy really cared, but it was something to break the monotony.

Sheriff McGrath halted, lifted his arm and pointed. Up ahead the land had been cleared years earlier by men seeking handy logs with which to erect the stores and homes of Puma Station. Some of those stumps stood nearly three feet out of the ground. A few trees, too spindly in those early days to be taken, had since that time grown to become impressive giants, shaggy and shadowy. Beneath such a tree a horse drowsed, tied to a low limb.

The possemen halted, sat staring, and looked at one another. "It's got to be a runaway kid," growled one man, showing disgust. "You know how much ground we've covered, only to end up just outside town a couple miles?"

"I know," said Tom McGrath. "That kid or grown man, he's up ahead in the shade of that big tree, and he knows darned well he stole that horse, which means he'll also know what's likely to happen when we ride up on him."

The men studied their terrain. There were no other trees close to the one under which the horse stood. But the ground was uneven, flinty and broken; men really wishing to could approach close enough, providing they were willing to creep and crawl to do it, and of course providing the horse thief up there didn't turn out to be a marksman with the eyesight of a hawk.

McGrath wordlessly led them over eastward where the arroyos were deeper, the little flowing hillocks brushier and in some places higher. Here they left their horses and began the long stalk. The sun was reddening now, dropping steadily down the western sky. Daylight would last another couple of hours, perhaps, and after that the Montana night would close in fast—and cold.

The men didn't bring along their carbines; only six-guns. They would get close enough. As a matter of fact, until the drowsing horse was roused, perhaps by their scent, that lumpy human shape up there on the ground didn't even stir, even though they sometimes boldly crawled straight over open places.

The horse saw them coming finally and acted uneasy, which is what awakened the thief on the shady ground.

At once an arm reached for and took hold of a battered old Winchester saddle-gun for which, evidently, the rustler had no scabbard.

As soon as the pale barrel began to swing, McGrath and his possemen became perfectly flat and perfectly motionless. They were within six-gun range now. They could have opened up on the thief, killing him with scarcely any genuine peril to themselves. But they didn't; not one of them was a timid man, or a cold-blooded one. Anyway, as McGrath himself said afterward, it got to be a sort of challenge to the men to see if they couldn't slip around and catch the horse thief by hand.

McGrath had the advantage; he could trace out an east-west erosion ditch all the way around the south side of the uneasy horse and emerge a short distance behind the pine tree. From there he had only to keep the big old tree between himself and the carbine, make no noise at all, and slip up until he was on the west side of the old pine, the horse thief on around in the opposite direction.

McGrath did that. He took lots of time doing it, too; the carbine looked as though it had seen plenty of service. The sun was even darker red by the time the possemen had stopped moving to see how McGrath came out of all this. They were still ringing the horse thief on all sides. They'd kill him in an instant if he detected McGrath and swung to fire behind the tree.

But that didn't happen, either, which naturally made the possemen feel a deepening contempt for this particular rustler.

McGrath covered the last thirty feet feeling his way with each hand before moving. He got right up to the back of the pine, straightened up onto his knees, drew his six-gun, didn't cock it just yet, and hesitated a moment to let his breathing settle a mite. Then McGrath stood straight up, stepped around and pushed his pistol into the back of the horse thief sitting on the ground, carbine in one hand, yawning mightily.

It was like touching a coiled snake. The horse thief whirled up off the ground in a spinning movement. Tom swung, striking the Winchester aside with his gun barrel, and lashing out hard with his balled-up big left fist. The horse thief's hat flew off, the gun was flung away, the crack of that solid strike sounded like a distant gunshot. The horse thief fell in a little heap at Tom McGrath's feet, and he stood there looking horrified.

"Hey," called a restless posseman. "All right to come on in, Tom; did you knock him out?"

McGrath knelt, rolled the horse thief onto his back and pushed back his own hat as he answered, "Yeah. Come on in. Only it isn't a *he*—it's a *she!*"

The possemen ran forward, not quite believing him until they saw for themselves. It was a "she" all right. No man living ever filled out a shirt like that, nor had long, wavy, reddish hair like that. The possemen looked from the unconscious girl to one another; in all their speculations, they'd never once imagined anything like this.

CHAPTER 2

THEY DECIDED AMONG THEMSELVES ON THE WAY BACK toward Puma Station to say nothing of the affair. No one knew there'd been a horse theft, anyway. At least until old Otto Meuller rode in the following day to get his horse and also to shoot off his mouth, probably over at Jim Astor's card room and saloon—called Angel's Roost for some inappropriate reason—folks wouldn't know.

As Tom McGrath said, he'd need a little time to find out who their pretty little prisoner was, and why she'd stolen the horse, and where she'd come from and where the devil she'd thought she was going.

The men were solicitous on the ride into town. Although it was dark before they got there, and they were helpful, even clumsily gallant, their horse thief wouldn't talk to them. She wouldn't even look at them if she could avoid doing it. She had a lump the size of a goose egg on the right side of her jaw.

But they'd been correct at least on two counts; she was dog-tired, and she didn't know the country at all. When she looked ahead and saw the lights of Puma Station after only a short ride from where she'd been captured, her astonishment was plain to them all.

Jack Kirk, the Two-Bar cowboy, sadly shook his head and, as though the girl weren't along, said, "Tom, what's the penalty for horse-stealin'; I mean for females?"

McGrath scowled at Kirk, saying nothing. He'd had Jack Kirk out on manhunts before, but this was the first time he'd ever seen the Two-Bar rider act so callous. It nettled McGrath.

Kirk seemed not to notice. He said, "Well, maybe we've never had a female horse thief in the Puma Station country before, but I'll bet you they know how to sentence 'em down in Butte."

"She won't be tried in Butte," a man growled at Kirk. "An' all of a sudden you're developin' a big mouth, Jack."

Again Kirk was untouched. "Why won't she?" he

demanded. "We got no means for tryin' females in Puma Station."

"A horse thief," stated the third posseman pragmatically, "is a horse thief. Don't make a particle of difference if she's a he or a she."

Kirk wagged his head. "You're jokin'," he answered. "It'll make all kinds of difference. You couldn't get a jury in this country who'd send a girl over the road for ten years for stealin' old Meuller's, horse. She'll have to be tried and sentenced down in Butte."

Tom McGrath began to suspect what Kirk was up to. He watched the girl. When Jack mentioned the ten-year prison sentence, her dark blue eyes widened noticeably. Her chin trembled. Sheriff McGrath sided with Jack Kirk, saying ten years was a light sentence considering that if a citizen-posse had caught her, they would have hanged her to the old pine tree back there. The other men looked shocked and embarrassed. They hadn't yet suspected what McGrath and Jack Kirk were up to. One of them fixed Tom with a dour glare and growled at him.

"She needs some grub, some hot coffee, and a place to sleep before you fellers start to bully her."

Puma Station was looming close by in the advancing night. There was a raw chill to the darkness now. Their prisoner shivered, and a cowboy draped his blanket-coat across her shoulders. She didn't resist, but neither did she look around to thank him. The coat made her look more forlorn than ever; its sleeves hung past her hands, and its shoulders drooped. She wasn't very large, although she was compactly, sturdily built. Neither was she very old; perhaps not more than twenty, if that old. And with those dark blue eyes, that heavy mane of wavy reddish hair, and a little tipped-up nose, she was attractive. In something besides faded old work trousers and an equally disreputable old woolen shirt, she would have been downright pretty. Even in those palpable cast-offs, she made them all very conscious that she was a girl.

McGrath said, "Miss, you can keep quiet as a clam as long as you like. But sooner or later you're going to have to tell your story. Stealing horses is a serious crime. If you won't talk in your own defense, you're darned likely to get the book thrown at you."

She looked over at Tom, studied his face a long moment, then said, "No one is going to send me to prison, Sheriff. No one."

The men gazed at her. She was defiant, which anyone else would have been under the circumstances, but she was also pale with fear and weariness. Jack Kirk said, "Lady, because you're you, it's possible you won't get ten years. But you're sure wrong about one thing: You're lookin' prison right in the face, whether you believe it or not."

She turned on Jack like an angered cat, spitting out her words. "You'll get buried if you try it, cowboy. You'll get buried before I'm even tried. Wait and see."

Kirk raised his eyes with a look of slow, hard triumph toward Tom McGrath. He'd finally stung her into the kind of talk he'd been trying to get out of her. McGrath nodded softly and put a skeptical gaze back upon the girl.

"Mind telling me your name?" he asked, and she flared up at Tom as she'd flared out at Jack Kirk.

"I mind. I'll tell you nothing, absolutely nothing."

Tom wasn't surprised. After all, this girl was intelligent. He hadn't really thought she'd be so fearful after all that talk of prison sentences that she'd break down blubbering. He looked over to where the back sheds and shanties east of town were coming up to meet them, and angled round so as to cross Main Street far south in the dark, and approach his jailhouse from a direction no one would be watching, now that it was fully dark.

None of them said anything more. They got around behind the jailhouse, tied their horses and took the girl inside through the back-alley door. Tom lit the lamp, hung it from its overhead hook, and went back to the sheet-iron stove to build a fire and put his little graniteware coffeepot on. It was cold inside as well as outside.

The man who'd given his coat to the girl strolled over to place his back to the crackling little stove and soberly regarded their prisoner. In a good light she looked even younger and smaller. She also looked husky and durable.

Jack Kirk pushed forth a chair, but she ignored it and stood across from the desk, staring with bitter defiance at them. That made the men feel somehow less manly; after all, there were four of them, heavily armed, big and lanky and rough, facing one small, stocky girl with a tipped-up nose and thick reddish hair.

The cowboy back by the stove shook his head and muttered under his breath, "I'm hungry an' thirsty. Tom, if you don't need me no more, I'll move along."

This man took another man with him. Eventually, only Jack Kirk and Sheriff McGrath were left to enjoy the warmth and the coffee, except for their defiant prisoner, who watched their every move without losing any of her spitting defiance. She only weakened when Jack brought back a cup of java, carefully ignoring her, and placed it upon the seat of the chair she'd refused when he'd gotten the thing for or her. Jack turned his back, went over by the roadside door and stood easily, sipping hot coffee.

She picked up the cup and drank. It wasn't good coffee, but it was hot and black.

Sheriff McGrath poked another scantling into the stove, stepped to his desk and said, "I'll fetch you some supper, miss, as soon as you tell me your name."

Undoubtedly it was a cruel thing to do, to take advantage of a starved girl like that, but Tom McGrath wouldn't have let her go hungry even if she'd refused to answer. He wasn't a merciless man; only a tough one.

She studied Tom closely for the second time, arriving at a secret estimate of him. "You make bad coffee," she exclaimed.

Tom grinned and nodded. "I do for a fact, miss. I also make heavy biscuits and boot-sole steaks." He strolled back to the stove, refilled his cup, held the pot up inquiringly toward Kirk, and Jack left the door to stroll over for a refill.

That was when she made her break. She flung down the tin cup, lunged for the door, and grasped the latch. Tom and the Two-Bar cowboy turned slowly to gaze over at her. The door wouldn't yield. She fought it desperately. Sheriff McGrath poured coffee into Jack Kirk's cup, set aside the pot and continued calmly to watch his prisoner fight for her freedom.

Jack Kirk finally said, "Miss, it's locked. Sheriff McGrath always locks it when he fetches home a criminal."

The girl flung around, her dark eyes blazing at them. She leaned both shoulders upon the door, fists clenched, hating the two men over by the stove with a depth of silent, deep rage they had to brace against as they returned her stare.

"You'll be the sorriest pair of men in this cow-camp town of yours for what you've done to me today," she hurled at them. "You'll want to get down on your knees before you've seen the last of me. I promise you that, both of you!"

Tom turned, refilled another tin cup, handed it to Jack who paced over and put it on the seat of the chair again for her, then went back and stood with his back to the merrily popping little stove again, all without a word. It was rather the way men might put out food for a wild animal.

She didn't go near the second cup of coffee. In fact, she didn't move away from the door until Tom McGrath, through with his coffee, said he'd go up the road to the café and fetch back a tray of grab for her, and walked toward the door. Then she backed off, watching Tom like a hawk. As he unlocked the panel and opened it, he gave Jack Kirk a look. The Two-Bar man nodded and went over to the door himself.

Sheriff McGrath had never before arrested a woman. He'd hurrahed a few out of Puma Station and had been thoroughly blessed out for his pains, but he'd never before had one in his jailhouse. While he was waiting for the café man to make up the tray, Tom leaned upon the empty counter speculating.

Jack Kirk had been right out there on the trail; no jury in the Puma Station country would find that pretty little rounded horse thief guilty of stealing Meuller's consarned horse. Tom wasn't even sure he'd want to make the charges against her, and even if he did, the townsmen and range men would make sly jokes about it as long as they remembered, which could be a very long time.

When the food came, Tom paid and walked back

down to his jailhouse. Kirk admitted him. Jack was smoking a cigarette, and so was the girl. That was something unusual, too; womenfolk didn't often smoke; or if they did, not in public, not in front of men who were total strangers to them. Not even in jailhouses.

Tom set the tray on his desk, locked the door again, jerked his head at Jack and returned to the pleasant warmth of the stove across the room. Not until then did he say, "Beef stew, miss, boiled cabbage and lemon pie."

She wanted to refuse; it showed in her expression toward them, but she'd been a long while without food, and the aroma was fragrantly strong. She went back by the desk and began to eat. She tried to do this decorously, but that good intention also failed. She wolfed the food down.

Sheriff McGrath went over to unlock his cell room, pass inside and heave back a clanking strap-steel cell door. Everything he did was in the manner of someone handling a wild animal. When he returned to the stove beside Jack Kirk, waiting for the girl to finish eating, he had the way open to drive her in and lock her up. Kirk looked at McGrath, making a rueful face.

She finished. There wasn't anything left, not even crumbs. Tom said, "Miss, before I lock you up for the night I want an honest answer from you: Are you carrying a hide-out gun in those clothes?"

She sidled away from the desk toward the roadside door again and didn't answer. She could see that waiting strap-steel cell in the adjoining room. It couldn't have been a very reassuring view under the most pleasant of circumstances for her. She flicked a look back toward Kirk and McGrath, saying nothing.

Tom shifted position. "I'm giving you a chance to be

honest with me," he said evenly. "Lady, you've got a couple of seconds to do that. Then I'm going to search you."

Her hands balled into small brown fists, ready to fight. Tom blew out a big breath, exasperation showing in his glance. He very clearly didn't want to search her. On the other hand, he had no intention of being shot in the back, either. As it was, he was bending over backwards just to avoid humiliating her, and if she lied to him—and was secretly armed—he might not live to reproach her.

"A simple question," he murmured, starting forward. "All you've got to do is give a simple answer. I don't want to tussle with you, miss."

Her back came to the log corner. She could retreat no further. She said, "No. No, I'm not armed. I only had that old cavbine you took away from me under the pine tree."

Tom halted. "Your word, miss?"

"Yes. My word."

McGrath motioned. "Then just walk through that door into the cell yonder and get some sleep. I'll leave the cell-room door open part way when I leave for the night."

"Why?" she asked swiftly, suspiciously.

Tom caught the implication and reddened. "So's the heat from the stove will keep your cell warm," he answered. "Now go on in there."

She obeyed. Tom followed, locked her in, then returned to the outer office where Jack Kirk was still smoking. Jack said, "She's not alone, Tom. You heard the threats and how she made 'em. She's got a friend somewhere around."

McGrath nodded, chose not to talk about this where

the girl could hear them, and said only, "Come on; I'll buy the first round up at Jim's bar."

They went out into the cold night. McGrath left the lamp burning inside, and he'd poked another log into the stove to keep his jailhouse warm all night. Now, as he securely bolted the door from outside, he said, "Why did it have to be a woman?"

Jack Kirk finally stomped out his smoke. "Why did it have to be right *pretty* one? Darned if I like the notion of bein' part of somethin' that's goin' to get her sent over the road, Tom. Neither did the other fellers; did you notice how they looked just before they ran out of the jailhouse?"

Tom had noticed. He led the way across the road and northward. There was one of those little icy winds blowing down out of Canada, and the night was black everywhere except right in the heart of Puma Station.

CHAPTER 3

BACK WHEN MOUNTAIN MEN AND PATHFINDERS WERE stealthily crossing Montana's high country, one eye peeled for grizzlies, the other eye peeled for Sioux, Cheyenne, Assinboin, any redskin at all because all were hostile in those times, an injured and starving man named Coulter had been attacked near a spit of pine forest not far from a respectable creek by an old bitch mountain lion. She may—probably did—have kittens close by in a punky old rotten tree; otherwise it was hard to believe she'd have attacked.

By all rights John Coulter should have lost that fight; he was weak and gaunt and cut and bruised. But he wanted to live badly enough to take on the old she-cat hand to hand, and although he suffered terribly from her claws—she never got a fang into him—he got a hammer-hold and strangled her to death.

It was almost two years later that Coulter came back at the head of a yeasty gathering of trappers, showed them the bones, explained about the fight, and made camp right there beside the creek where he'd strangled the old girl.

After that, when trappers, soldiers, Indian scouts, even the red men themselves, camped at this place, they called it by the name Coulter had given it: Puma.

That name was foreign to Montana. It was the south-western name for a mountain lion. In the north lands pumas were known as cougars, mountain lions, or just plain lions. They were also known as catamounts and varmints. Long after John Coulter stopped coming, long after the beaver trappers saw their market collapse, long after the last redskin buck had ridden away from that place, it was still called Puma.

There was a trading post there. Eventually the army cut a road to Puma. Years afterwards, when the army was no longer needed, stores blossomed and Puma became almost a town. What clinched it was the coming of the telegraph and stage lines. Every stage company had way stations, usually near some town where they'd be reasonably safe from outlaws and marauders. Puma became Puma Station, and by that time scarcely anyone was yet alive who remembered how itd gotten its first name.

The cattlemen who came to drift their herds over the endless, incredibly rich ranges didn't give a hoot in Hades where the name derived from. Neither did their range men. After a while, neither did anyone else, including Sheriff Tom McGrath, so when he showed up at the jailhouse the following morning with a tray of breakfast for his husky little prisoner, and she asked him without any preamble how the town had come by its unusual name, Tom was surprised.

He was also pleased. She didn't spit at him; didn't even make her grisly threats again. After he'd given her the food, he told her all he knew about the name of Puma Station, and since he'd also been in the Southwest, he also told her that puma was the Mexican word for a mountain lion.

She then asked him if he'd like to know her name. He said that he would; in fact, he'd promised himself to find it out today whether she told him or not, because he couldn't book her into his jail without some kind of a name.

"Ruth Hall, Sheriff McGrath."

He was pleased, not just that she'd changed so completely overnight, but that she was being friendly and co-operative. He wanted to say something reassuring but couldn't frame the right words. "Thanks," he told her. "Otto Meuller, who owns that horse you took, will be in today for a look at you, more'n likely. Beyond that, and asking you why you took that horse, where you came from and where you were going, I'll leave you alone."

"The horse," she said matter-of-factly, gazing out at McGrath, "replaced one I lost two days ago. That's the last time I had anything to eat until last night. I can't stop this Mister Meuller from wanting me prosecuted, I guess, Sheriff, but if you'll let me talk to him before he makes you do that, maybe I can convince him I only borrowed the animal."

Tom nodded down at her. He'd let her see Otto alone, but he knew something she didn't know. Otto Meuller would no more be talked out of something like this, even by a very pretty girl, than he'd give up being an overbearing, gruff, tough old-timer on the cow ranges.

"Where were you going?" he asked, and ran head on into his first stone wall.

She looked him straight in the eye when she answered. "I can't tell you that, Sheriff. If things don't work out for me, I'll tell you, but right now I can't."

Tom nodded, rebuked but unruffled. "Where were you coming from?"

"Idaho," she answered promptly. "Idaho by way of the Oregon Territory."

"And you lost your horse."

"I lost him," she answered quietly, dropping her head to resume eating. "Lost him with a bullet through his head."

McGrath looked around, found the nail keg he kept in his cell room and sat upon it. "You were waylaid?"

She didn't answer. The breakfast was nearly gone. She picked up the coffee cup, emptied it, put it down and said, "Sheriff, do you have the makings?"

He had, and he fished them out. Then he sat beyond the bars watching those small, strong hands professionally twist up a cigarette. H'ed seen women do that before, but he'd privately held them in low regard. He handed a match through to her also, waited until she'd lit up and exhaled, then said, "Miss Hall, who knocked the horse from under you, and how'd you get this far on foot?"

"Sheriff, I don't know who shot my horse. I didn't stay back there trying to find him. I ran as far and as long as I could, came upon a trapper's camp, stole that carbine you took from me, stole some jerky, and kept right on running. Oh, yes, I also stole these clothes."

She handed Tom back his makings through the bars and looked up into his eyes. "You're not what I expected, Sheriff; I thought, when someone finally ran me down, he'd be unshaven and smelly and leering."

Tom thoughtfully pocketed his tobacco sack and papers. "Why were you running, Miss Hall; what was there back in Idaho you wanted to get away from?"

"Men," she said, and spat the word out.

McGrath nodded and stood up, believing what he wished to believe—which happened to be very wrong. But then Tom McGrath was a bachelor, and a gallant one to boot. "I'll write it up," he said. "When you're finished I'll take away the tray." He turned to leave.

"Sheriff, who was that cowboy with you last night—the lean, blue-eyed one you called Jack?"

"Jack Kirk," he answered. "He rides for the Two-Bar outfit. I've used him on posses before. Why?"

"Is he the only deputy you have?"

"No. I don't have any regular deputies. I don't need them. Jack just happened to be in town when I heard someone had made off with a horse from the Meuller place, so I sort of dragooned him and those other men into riding with me to catch you."

She dropped her head again, and Tom went out to his desk to complete the booking. When he finished with that he pushed back his hat, leaned on the desk and stared at the wall. This was no common horse thief in a lot of ways. She was sharp and quick-witted and—somehow—he didn't like the change in her.

For one thing, it was too sudden. People's moods didn't change that much that quickly. There were shades to all colors; black didn't just turn into white overnight. Another thing he didn't like was the facility of her answers. He believed them. At least for now he believed them. But something was out of kilter; something was jarring on his nerves about those answers. In the first place, a girl didn't just jump on a horse and try to ride half across creation just because some men had leered at her. If that had been her sole reason for leaving Idaho, she wouldn't have had to go any farther than the Oregon Territory. In the second place, she was running too hard, too long, just to be trying to get away from bad memories.

He put aside his book and pencil, wrote two lines on a scrap of paper and sat awhile gazing at what he'd written. Then he straightened his hat, left the jailhouse for the telegraph office, and had those sentences transcribed on a yellow slip of paper for transmittal to Boise, the biggest town in Idaho. After that he returned to the jailhouse and waited for Otto Meuller's arrival. It wasn't a very long wait, for although he hadn't sent word out to Meuller he'd retrieved the stolen animal, he'd known for a certainty Otto would drive into town in the morning.

Meuller was a bull-necked man with small, pale eyes, fists the size of hams, and a disposition hard to describe. He had two buxom, flaxen-haired daughters and was a widower. He was a top cowman and had made money every year, even through the droughts and blizzards, but he ran his ranch as if it were an empire and he the king. He'd never been able to keep riders very long, not just because he refused to permit them anywhere near his daughters, but also because he was a violent, rock-fisted man who answered every argument with a challenge. The only thing that had kept him alive this long was the fact that Otto Meuller never wore guns.

When he stamped into the jailhouse and saw Sheriff McGrath poking wood into his sheet-iron stove, Meuller let out an indignant bellow: "How come you aren't out after that darned horse thief instead of warmin' yourself by the fire? Is that what we pay taxes for, I'd like to know!"

Tom turned and straightened up to his full height. He wasn't quite as tall as Meuller and he wasn't as broad, either, but they were quite different in build. Meuller was like a bull: massive, heavy, thick and powerful. Tom McGrath, who was only average height, was like oak. He was also one of those inexhaustible men who could go indefinitely on very little sleep and even less food. In a confrontation it would have been hard to decide where to place the betting money.

"I've got your blasted horse," he said to Meuller, "and I've got the rustler too." Tom pointed to the cell-room door. "Go on in. She's expecting you."

Otto blinked. "She . . . ?"

Tom didn't reply. He stepped to the door and pulled it open. Beyond, the girl sat in her cell gazing out at Otto Meuller. He saw her, too, of course, and stood as if he'd taken root. "That's the one?" he asked in a fading voice. "Tom, is that who stole my horse?"

McGrath inclined his head and went back over to tend his stove.

Meuller stood like a statue for a moment longer, then shuffled over to fill the cell room doorway, still staring. Tom heard the girl say something. He didn't hear any answer. Tom Astor, proprietor of Puma Station's most popular saloon, poked his head in the door to say there was a fight in the road outside. Tom crossed his office and went out into the fresh morning sunlight. Up in front of the general store two men were furiously pummeling each other while townsmen stood back watching, and several outraged wives were noisily urging the menfolk to intervene. So far, no one had even walked anywhere near.

The battlers were a pair of freighters. What had occasioned their brawl, aside from short tempers in the early morning, had been a simultaneous arrival out in front of the general store, where both had tried to beat the other to the parking place to unload their deliveries. As far as Tom McGrath was concerned it didn't amount to much. Unarmed men fighting were no novelty, and seldom were such fighters really dangerous. He didn't even break stride as he came in behind the two, who were so perfectly matched that neither could force the other to give an inch. They were swinging with all the ferocity and none of the cool science of men who really knew how to fight. Their shirts were torn, their faces lacerated, their lungs pumping like steam engines. It would have no doubt ended in a draw, but as Tom McGrath moved in he grabbed the nearest man, whose back was to him, whirled the big freighter and uncorked a blasting right fist.

The freighter probably was too groggy to see the fist coming. At any rate, he made no move to avoid it. The blow popped like a bull-whip. The freighter dropped straight down, rolled off the plank walk and didn't move again.

Tom then headed straight for the other one. This time, though, his adversary knew what was coming and tried to beat McGrath to the punch. All he really did was miss with a big looping blow and catch Toms left in the belly, which doubled him over, then caught a vicious rabbit punch at the base of the skull that dropped him face first at Sheriff McGrath's boots.

Jim Astor, the saloon man, gave a long, silent whistle. He was used to brawls, but not two against one, both bigger men than Tom McGrath, and not when the two had been taken out so swiftly and expertly. He said to an aproned clerk in the doorway of the general store: "Fetch a big bucket of water." Then he looked at Sheriff McGrath with great respect. "You goin' to look them up, Tom?"

McGrath shook his head. "When they come around, tell them I said to unload their wares and get out of Puma Station."

Across the road, Otto Meuller was standing outside the jailhouse at the hitch-rack, looking at the crowd up by the general store. Tom flexed his knuckles, shouldered through and headed back toward his office. When he got close enough to see the expression on Meuller's face, it shocked him. He'd known Otto Meuller a long while, and he'd never before seen Otto looking as he looked now.

"What's wrong?" he asked.

Meuller looked straight at McGrath, drew up off the tie-rack and walked around it to the side of his horse without uttering a sound.

Tom grabbed Meuller's arm. "Where do you think you're going, Otto? You've got to come inside and sign that complaint."

Meuller draw his arm away. "Sign nothing," he muttered in his deep, gruff voice. "I'll sign nothing."

"Otto. Hang it all, she stole your horse!"

Meuller blinked like a man in a trance. "Stole my horse? No. I give her that horse. She didn't steal the horse, Tom. I give it to her."

McGrath stepped back and watched the cowman hoist himself up across leather, shorten his reins, turn and ride south at a slow walk, oblivious even to the crowd up in front of the general store where the two drenched freighters were struggling to arise.

Finally, Tom turned and went into his jailhouse. Through the cell room door he saw Ruth Hall standing against the front of her cell smiling out at him. "Open the door, Sheriff," she said sweetly.

CHAPTER 4

BUT TOM MCGRATH, GALLANT THOUGH HE COULD BE and definitely had been toward his prisoner, was also a tough man. He not only refused to release Ruth Hall; he didn't even act very pleasant when he demanded to know what she'd said to Otto Meuller.

Her smile stayed put but it became strained as she looked up into the lawman's face. "I explained what happened to me; why I had take his horse—someone's horse—and it just happened to be his buildings I saw as I came wandering through the night."

"Sure," said Tom sarcastically and softly. "So Otto Meuller, who wouldn't give a man dying of thirst the sweat off his brow, decided on nothing more than your explanation—and your smile—to give you the horse and to refuse to agree to sign a complaint against you. Lady, you're a doggoned liar."

She still smiled; only now it was hard, fierce little smile as she told him he would have to release her. He leaned on the cell room wall stonily regarding her. "I made a bad mistake," he said finally. "I reckon no man likes being made a fool of." He turned to leave the cell room. She called his name sharply.

"Sheriff McGrath! You *have* to release me."

He didn't even turn around until he was through, and then only long enough to lock the cell room door over her vehement protests. The door was a heavy slab of fire-hardened oak; some predecessor lawman's ingenious answer to the lack of steel for this particular door. But it didn't quite muffle Ruth Hall's fury. Tom McGrath stood on the far side of it, listening with an experienced man's critical attention, but she didn't swear; she used perfectly acceptable words, although in such an eloquent manner that he had, despite himself, to admire her ingenuity.

Then he left the jailhouse, heading out back to where he kept two saddle horses in a lean-to with an attached pole corral. He didn't quite make it. Jack Kirk came down the sidewalk to intercept him before he made it to the corner of his building.

Jack said, "Hey, Tom; on my way into town I ran into Nels Bagley. He told me somethin' I wanted to verify with you. He said Meuller went past him on the range heading for home, and Otto told Nels he'd given this girl horse thief that horse she stole. Nelson said Otto told him you'd tried to get him to sign a complaint and help you prosecute her, her that he refused because she never stole that horse at all. He done gave it to her." Kirk looked almost ready to smile at the palpable falseness of this, but as he studied the sheriff's gloomy, unsmiling face, his grin sort of slipped sideways even as Jack Kirk's gaze turned a little skeptical, a little baffled.

"Otto told the truth," said McGrath. He then related what had happened and how it had happened. He also said that he was right that instant on his way out back to saddle up and go catch Otto; sweat out of him what the cussed girl had said to put the fear of God into him.

Kirk strolled along with Tom McGrath, lost in thought. He didn't even tell Tom until the sheriff was going in after one of his horses that the ride would be fruitless because Nelson Bagley had told Jack that Otto hadn't been heading for home; he'd been heading upcountry on a big circuit of his range.

McGrath then went to the corral stringers nearest Kirk, which happened to be in the shade, and made himself a smoke, also delivering himself of a pithy comment.

"I don't need Otto anyway," he stated. "I can guess pretty close what she told him. Only I can't guess *who* or *where.* It's not that complicated when an old cuss like Meuller's involved. Only a couple of things could really frighten him. I wasn't smart enough to figure it out at first, or I'd have made him stay in town."

Jack fell to making a smoke. "*Made* him?" he said. "*Made* that square-headed cuss do something?" Jack lit up, shaking his head. "I've been on roundups with Otto Meuller; nobody *makes* him do anything. Believe me about that, Tom."

"Jack, you're talking through your hat. I just told you what that girl did to him. Otto's the kind who'd want the Angel Gabriel shot, hanged or imprisoned for stealing one of his horses. But Ruth Hall didn't just back him down, boy; she *made* him refuse to prosecute. She *made* him give her the horse she stole from him. Now you tell me again no one can budge Otto Meuller."

Kirk nodded and smoked, and eventually said, "I can guess what she told the old cuss. We figured out last night she wasn't in the country alone. All right; she told Otto some gunfighter was skulkin' around, and if Otto prosecuted her, this here gunfighter friend of hers would maybe shoot Meuller."

"I like my version better," grumbled the sheriff. "She told Otto her gunman friend would kill Otto's *daughters.*"

The two husky men stood gazing at one another; then Jack Kirk conceded again, nodding his head to indicate that this was so. He ran a hand through his light hair. "If you got this much of it figured out, Tom," he said, "what's behind it?"

McGrath had no ready answer, but he'd been a lawman long enough to have some opinions. "Who knows?" he mumbled. "Anyway, there's a lot more than we can guess about."

"Like what, for instance?"

"Her horse. Did someone really shoot the thing out from under her like she told me, or what? And why she's on the run; not to mention if she's really from Idaho, why she ran so cussed far. A lot of little things like that. Those old clothes she's wearing. Who this friend of hers is."

"Maybe he's no friend at all, Tom. Maybe he's someone she's runnin' from," exclaimed Jack Kirk. "Only if that was so, why would she say—or at least hint—that he'd settle Otto's hash for him if he gave her trouble?"

Tom McGrath didn't have the answers, and he'd speculated himself plumb out. He turned to watch his horses go stand under their overhang in front of the manger and nibble indifferently at hay. The trouble with an affair like this, of course, was that nibbling around the edges only whetted a man's appetite to know more, and meanwhile, whatever criminal acts were involved remained unpunished and very likely to be added to.

"I've got to find this gunfighter friend of hers—or whatever he is."

"How? This is a big country, Tom, with lots of miles to it, and this time of year lots of strangers work the ranges and towns. In wintertime it'd be fairly easy. Aren't any strangers around then."

"Well," stated McGrath, "he hasn't been in the

country more than a day or two, and he'll be interested in the girl, asking questions and the like. How about you helping, Jack?"

Kirk was perfectly willing. He just wasn't very enthusiastic about their success. "Sure; I'll nose around in town and out on the range. But I'll be more surprised than you will if that turns up anything."

"There's another way," said the lawman quietly, still gazing over to where his horses were loafing. "I can wait for him to come to me. If he really wants her out, he'll have to do that, providing I keep her locked up."

Jack asked the inevitable question: "Why? Sure she's pretty. Built like a purebred heifer, too, an' I reckon that's enough reason for him to want to rescue her. Only as I see it, Tom, this isn't his game as much as it's her game."

McGrath agreed that a man would want to help Ruth Hall if for no other reason than that she appealed powerfully to every masculine instinct. But it wasn't her prowler friend McGrath was as much interested in as it was the reason for these two to be in his country acting as they were. But he reasoned thoughtfully and it still came out the same way. "I don't even hope to know all the answers, Jack. I just aim to find them out by keeping her locked up until she either tells me what this is all about, or *he* shows up."

Kirk scowled. "'You ever think, Tom, he might show up all of a sudden; maybe from behind a building as you're walkin' by, or in an alleyway?"

"Sure. That's the same chance I've been takin' since I first signed on as old Jamieson's deputy. I'll have to take the same chance now." McGrath walked outside the corral, secured the gate and said he had some work to do; he also reiterated his desire that Jack Kirk see what he could detect among the range riders. They broke up after that, Kirk ambling rather aimlessly up toward the Angel's Roost Saloon, Sheriff McGrath heading briskly over to the telegraph office. His wire to Boise'd had plenty of time to get there and be answered.

It had been. The telegrapher, a long-faced individual with sad eyes and a perpetual sniffle, handed over the reply with a perfectly impassive face.

No one in Boise's U.S. marshal's office had ever heard of Ruth Hall. If she was wanted in Idaho, no such information had come to the marshal's office from any of the counties, nor from any federal agency either. The telegram closed with a weak hint of encouragement by saying inquiries would at once be undertaken, and if anything worth-while turned up, Sheriff McGrath would be notified.

Tom went over to get something to eat. Afterwards he took another tray of grub to Ruth at the jailhouse and walked into another storm of non-profane but quite adequate abuse. In the end, though, she ran out either of expletives or of breath, and accepted the meal through the cell door, placed it upon a bench and watched grave Sheriff McGrath relock her door, draw up his rickety horseshoe-nail keg, sit down and cross his arms, eying her.

"I dont know which is worse," she said candidly, "you sitting out there like Sitting Bull staring at me, or bringing food in here when you've no right to keep me here at all."

"Well," he laconically commented, "I could forget to fetch food, and still keep you in here."

"But, Sheriff, there's no signed complaint. How can you legally do this?"

"It's right easy, Ruth. I just don't unlock that door."

Her eyes flashed with exasperation at his seeming imperturbability. "Even though it's not legal?" she demanded.

"It's legal, Ruth. You're in protective custody."

"What's that?"

Tom smiled thinly, looking her straight in the eye. "Well, it's a term we use in my trade which means we're protectin' someone from someone else. In other words, maybe that feller you stole the clothes and the old carbine from is huntin' you. Maybe the man who knocked the horse from beneath you is still after you, too. Or maybe, Ruth, this friend you're always hinting will make mincemeat out of me for holding you in here might have a change of heart and decide to liquidate you."

She crossed to the bunk, sat down, looked thoughtfully at the tray of food and finally turned. "In other words, Sheriff, you're holding me here because you want to."

He nodded, still sitting there like an Indian, arms folded across his deep chest, face relaxed and impassive. He gave the impression of having all the time in the world. He also gave the impression of being a man unlikely to be moved by appeals, threats, or even teasing smiles.

She said, "McGrath, I wish I could be a man for fifteen minutes—outside this cell."

That tickled him. He arose, smiling. "I like it a sight better the way you are, Ruth." He stepped to the cell-room doorway and kept on smiling at her. "The question seems to be—which one of us can afford the longest wait. I'll tell you this much: I'm in office for three more years."

"What will it take?" she asked softly.

"Just the truth. That's all; just the simple, unembroidered truth."

She sighed, turned and started eating. Tom McGrath

might as well not have existed. He still stood there, though, and there was something he liked very much about Ruth Hall: she didn't lie.

She could have concocted some cock and bull story the day before when he'd interrogated her, but she hadn't. All she'd said was that she couldn't tell Tom anything. Now again, she refused to lie and simply turned resigned.

McGrath couldn't have denied it if he'd wanted to: he liked her. Probably because she was a full woman in every sense, but he liked her for what she inherently was as well. A man in McGrath's position heard lies every day of his life; sometimes clever lies, but more often stupid, clumsy ones. He had a strong aversion to lies and liars. Like most uncomplicated men, Tom had a lot of faith in simple virtues.

"Lady," he told her, "I'm even enough of a hick to believe you're using your correct name. That's pretty naïve, isn't it?"

She went right on eating, ignoring him with a frosty, total indifference. He thought of something, fished out his makings and tossed them through the bars where they landed on the cot beside her; then he stepped through the cell-room door and locked it, unaware that she'd finally lifted her eyes to him.

The balance of the day went by calmly and uneventfully. Jack Kirk was gone when McGrath went up to Jim Astor's place in the evening for a little poker and a few nightcaps.

No one asked questions, although by now it was no secret he had a female horse thief in his jailhouse. Probably that was because the riders and townsmen hadn't seen Ruth Hall and assumed, because women outlaws were usually notoriously horsy and sleazy, she was just another old crow.

Jim Astor, who owned the Angel's Roost, brought over a free drink for McGrath. He hadn't recovered yet from the way Tom had put down those two big, rough freighters with two punches. Jim was a wispy, grey-headed man who looked much older than his probable thirty-five or so years. He was a smiling man, who could joke one minute and be rough the next minute. He was strictly a man's man. When others had approached him about importing dancing girls, he'd turned thumbs down, saying that his Angel's Roost was a saloon, not a hog ranch; if the boys wanted something besides poker and liquor, they'd have to open their own saloon.

Tom and Jim were friends without being intimate. They'd had their little disputes, too, which was inevitable, considering their opposite vocational leanings. Still and all, Jim Astor admired the sheriff; even when he was howling the loudest about Tom interfering with his drunken and troublesome customers, he was ready to back Tom up if a free-for-all started. As he'd once told some cowmen: "Puma Crossing never had a *real* lawman before, and I think we should support him one hundred per cent, whether we always agree with him or not."

And that was the way it was in Puma Crossing.

CHAPTER 5

MCGRATH HAD HAD RUTH HALL IN JAIL SIX DAYS AND was fast weakening; after all, his jailhouse hadn't been built with women in mind. It had its definite drawbacks. Furthermore, she never ceased nagging at him to be set free, although she never offered to answer his questions either. In Tom's opinion, she was just plain exasperatingly stubborn, like a Missouri mule.

He was on the verge of turning her loose. Jack hadn't turned up anything among the range men. Nothing had occurred around town, and although McGrath had gone out to try and pry something out of Otto Meuller twice, all he'd gotten was a dirty look and some growls of defiance.

So, regardless of his bluff concerning keeping her locked up indefinitely, he was very close to admitting defeat the day old horse-face-and-sad-eyes the telegrapher ambled over to hand Tom silently one of those yellow pieces of paper his telegraph company used for official communications. That changed everything.

The U.S. marshal's office over in Boise, Idaho, came up with something shocking. Ruth Hall was known to have come from a small town along the Snake River, where she'd been seen in the company of one Jay Adams. There was more, but Tom closed his office door, tossed the telegram atop his desk, and went over to stoke up a little fire under his coffeepot before reading it.

Jay Adams was a notorious outlaw; perhaps *the* most notorious train and bank robber in the entire Far West. He'd dropped from sight a year back after a particularly successful holdup in Nebraska, which was so far out of his normal territory no one had actually believed he'd masterminded that train robbery until one of his men had been shot in Laramie, and on his death bed had unburdened what passed for his soul.

And Jay Adams could well hole up over in the Snake River country for a year; he'd gotten not less than a hundred thousand dollars from that Nebraska robbery. It had caused a tremendous furor, not only in the West, but also as far East as the nation's capital. Pinkerton detectives had been on the trail. So had army intelligence teams, and every bounty hunter in the trade, for the reward was posted as one third of the stolen money.

Six months after that stunning holdup, the manhunters had trickled back to the towns without a single lead as to the whereabouts of Jay Adams. Several of the renegades who'd been with Adams over in Nebraska had been picked up—including the one who'd been salted down in Laramie—and invariably they were broke when apprehended. But every one of them had said Adams had fairly given him his share.

That, as McGrath had reason to know, amounted to something like half the stolen money, or perhaps slightly less than half. Jay Adams then still had in the neighborhood of fifty thousand dollars in cold, hard cash.

Tom waited for the coffee to heat, then took a cup of the stuff back to his desk, sat down and finished reading the telegram. The U.S. marshal in Boise was notifying the U.S. marshal in Helena, Montana Territory, to go at once to Puma Station and see if this was indeed the same girl, and get out of her the information as to where Jay Adams now was.

Tom drank his coffee, turned and gazed at the closed cell-room door. Instead of a paltry horse thief, he had a national celebrity. At least it looked like it from the wording of that telegram. The Boise authorities were excited, that much was plain. They were also about to cause Tom McGrath to be visited by some of those quietly efficient, very deadly, invariably suspicious and distrustful U.S. deputy lawmen.

He returned to the stove, poured a fresh cup of warmed-over java and took it into the cell room, where he pushed it through the bars. "We've got some talking to do," he said, as Ruth took the java and went back to her bunk to sit and sip and study McGrath's craggy features.

"There's nothing to say, Tom. We've been over it all before. You're holding me illegally. When I finally get out I'm going to write the governor and the—"

"Cut it out," McGrath growled, looking around for his keg and drawing it up close to the bars. "Ruth, why is Jay Adams after you?"

She didn't betray anything on her face, but her hand shook, spilling a little of the coffee. She looked quickly away while she mopped it up with a blue bandanna. "I don't know what you're talking about," she muttered.

"Don't lie," he said sternly. "If you lie to me now, you're going to destroy something."

She glanced swiftly out at him, then glanced just as quickly away again. Evidently what he'd just said had some particular meaning for her.

"Ruth, I got a telegram from the Boise office of the U.S. marshal. He told me about you and Jay Adams over in the Snake River country. And I've read the newspapers the last couple of years. I know Adams still has most of that money from the mail coach over in Nebraska. All I want you to tell me now is why he's after you? Did you run off with the fifty thousand?"

She sipped coffee and ignored him. He hitched his keg a little closer, his face grim. Outside, some boys ran by, making fluting sounds in the golden day. Neither of them heeded this or any other sound. She avoided looking at him and sat stiffly inside her cell. Tom's unsteady horseshoe-nail keg creaked under his solid heft while he stonily looked in at her.

"Federal deputies are on their way down here," he told her. "You won't stand the chance of a snowball in hell once they get hold of you."

"Are you going to hand me over to them?" she quickly asked.

He straightened back. "What choice have I? They'll have federal warrants from Idaho, from Nebraska, who knows from where else? Jay Adams isn't some two-bit gunfighter, Ruth; even as far off as Washington, D.C. they want him. He's probably robbed more banks and trains and gotten away with more money than anyone, including the James boys and their kinsmen the Youngers. Girl, next to Jay Adams, Jesse James was a marble-shooter. I'll have to hand you over."

She put down the tin cup and slumped a little, studying the hard earthen floor at her feet. "They'll never get me to Helena, Tom, if that's where they'll try to take me."

"That's what I'm drivin' at, Ruth," McGrath said earnestly. "Give me something to use; tell me why he's after you and where he is."

"No," she whispered. "I couldn't do that."

He exploded. "Why not? Listen to me, Ruth. If you've got some crazy notion of loyalty to him, or think that you owe him something, just balance those things against what he'll do if he catches you."

"I know what he'll do. I just told you, Tom; the federal lawmen won't get me fifty miles from Puma Station."

"Then what in the—!"

She turned swiftly. "Tom, you don't understand at all. I know you've been trying to be helpful in your own way. I like you for it. I really do. You've treated me better, even when I was a prisoner in your jailhouse, than most men've treated me when I was free. But I can't help you kill Jay Adams. I just can't."

He sat a moment regarding her though the bars. "Well," he eventually said, softly and slowly, "if you love a man, why then I reckon you feel you got a duty. But, Ruth, not *this* man. He dynamited that mail coach in Nebraska, killing three clerks and a guard inside. In Texas he and Sam Bass blew up a bank with more dynamite than they'd need just to burst a safe—and when the ceiling fell in, six folks died who were asleep in the hotel upstairs. I could keep this up for half an hour."

She rose, walked to the back wall of her cell and kept her back to him. She looked forlorn and abandoned in those old patched clothes, which had belonged to a man who'd been much taller and less compact. Tom slapped his legs, stood up and stamped out into his office, locking the cell-room door behind him.

He wanted a smoke but didn't have any makings, so he went out into the pleasant hot daylight, paced over to the general store and bought a sack of Durham. As he returned to the roadway, he knew what she'd told Otto Meuller. In the light of what else he knew, he now understood very easily why Otto had turned tail. To mention the name "Jay Adams" in connection with revenge was enough to turn anyone pale.

He built his smoke, leaned upon an overhang post outside the general store, balefully gazing down at the jailhouse, and failed to see Jack Kirk until the youthful range rider halted over to his right, stepped off and looped his reins at the general store's hitch-rack.

"Well," said Jack, strolling over, "I've got news for you, Tom."

McGrath turned. “That’ll make two of us,” he mumbled, dropping his smoke to stamp on it. “Let’s hear it.”

“There’s a butchered Two- Bar steer on the west range.”

McGrath’s suspicions instantly leaped upon this. “A yearling or less?”

Kirk shook his head. “A two-year-old, Sheriff. A critter that stood in live weight right at fourteen hundred pounds.”

Kirk stood there watching the significance of this soak in. McGrath’s brows rolled down slightly. “I’ll ask a darn fool question,” he murmured. “Why such a big one?”

“Get saddled up and I’ll show you why,” replied the cowboy. “And the Two-Bar range boss is mad enough to eat coal and spit fire.” Kirk smiled thinly. “Only he’s not quite mad enough to take the crew and go after ’em.” Jack tipped his hat to a pretty girl who swept past into the general store, and followed her with interested eyes as he went on speaking. “They ate most of one rump, Sheriff, and carved up that meat to take with ’em. About all they left was the stuff no one eats much anyway—neck, brisket, shoulders.”

McGrath kept waiting for Jack to turn back and face him again. When that eventually happened, he said, “Had to be more’n just a couple of them.”

Jack nodded. “That’s why the range boss didn’t want to go rushin’ after them. Seems rustlers in ones and twos are his meat, but more’n that, an’ he gets indignant for the law to do something. So he sent me in to tell you there’s got to be something done.” Kirk’s spontaneous, sudden smile broke out. “Like maybe callin’ in the army or somethin’.”

Tom said, "You ever hear of a man named Jay Adams?"

Kirk's smile dropped away. He looked straight at McGrath with a gradual expression of dawning wonder across his face. "You mean—here? You mean—that's who her friend is?"

"Yeah."

"Holy mackerel. . . . He ate the Two-Bar steer?"

"Who else travels with a gang?"

"Well, I don't know. Maybe some of those hide-out redskins who're always jumping back and forth from the Canadian forests down here, then running back again. Or maybe—"

"See any moccasin tracks around that steer, Jack?"

"Well, uh, no. Matter of fact, the sign was all boots and shod horses."

"Then it wasn't redskins, was it?"

Kirk scratched the back of his neck. "I reckon it wasn't. . . . You sure about this Jay Adams tale, Tom?"

"I haven't seen the man, if that's what you mean, but I've got a telegram over at the jailhouse that'll be an eye-opener for you."

"Yeah. Him and the girl. I understand. Only . . ."

Jack looked squint-eyed over at the jailhouse, then down at the dust in front of him.

McGrath saw the consternation and confusion mounting in his friend. "All the same. I'll go get my horse and meet you here in a few minutes. I'd like to see that steer too." He stepped down into the morning sunlight. "Keep out of Jim's place," he growled, and went striding off.

Kirk didn't go near the Angel's Roost Saloon. He ambled into the general store for some tobacco, and afterwards stood in the same overhang shade McGrath had been standing in, doing the identical thing Tom had; he made a slow smoke, lit it, and put a quiet, curious glance over upon the front of the jailhouse.

McGrath wasn't gone long. When he and Jack were riding up out of town, the cowboy said, "Hey, I just thought of something. Isn't there some kind of big reward out for Adams?"

McGrath turned rueful about that. "Sure; more'n you'n I'll make in ten years. But getting Adams might be the hard way to earn that kind of money, especially if he's around here in the mountains somewhere with another of those gangs of outlaws he recruits whenever he's on the prowl."

They rode for a while through an empty, still and silent world of curling grass, little land swells, and occasionally a tree or some flourishing, spiny brush. They saw cattle, too, far off, and eventually detected the rising dust behind a band of running horses.

Where the Two-Bar steer had been killed was an ideal place for it; there was a little creek within fifty feet of the carcass, and the slaughter ground was tucked around a bend in the low, thick hill which jutted down from the rougher hills farther back.

There were plenty of tracks. Even granting, as Tom said, the Two-Bar riding crew had trampled the place, there was still plenty of sign of others. For instance, at creek-side, they found where men had knelt to wash and drink. They also found a stone ring, farther up the hidden gulch, where, supper had been cooked. Tom pointed to the size of the fire and looked grim.

They even found cigarette stubs, and according to McGrath, that was the best clue they had yet as to the size of the gang. "Ten, at least. Maybe I'm off one or two men, but if there are only eight of 'em, I'll be switched."

"More'n likely," opined Jack Kirk regarding the deftly carved up carcass, "you're off a couple in the other direction. Say, does this Adams always travel with an army?"

"You've seen the same newspapers I have," mumbled McGrath, returning to his horse, satisfied with all he'd seen. "He can afford to pay good, and when he lets it be known he's ready to ride, they say gunmen and renegades come from all over to ride with him."

Kirk also returned to his horse. They mounted up, cast a last look around; then Jack said, "They didn't go much out of their way to hide the sign, did they?"

"Adams doesn't believe in skulkin', the way I've heard it. He bellows like a bay wolf and dares anyone to trap him. Let's get on back."

CHAPTER 6

TOM MCGRATH WASN'T WORRIED ABOUT JACK KIRK saying anything. Jack was one of those blithe spirits who had a secret side. After they parted in town, Jack to return to Two-Bar and assure his range boss Sheriff McGrath would get right on the trail of the rustlers, Tom went into his cell room to tell Ruth where he'd been and what he'd seen. At first sight of him she started to say something about her tardy supper, but after he'd been speaking for a moment, her expression altered, her shoulders sank, and she evidently forgot all about her hunger.

When McGrath was finished he looked in at her, not unkindly, and said, "Ten men after one girl. Ruth, if you didn't make off with his cache, I don't understand it.

Oh, you're pretty as a calendar picture all right. But—*ten men* just to get you back?"

She could have shot a retort back about that, but she didn't. In fact, she didn't say anything until someone stamped into the outer office from the roadway. Then, as Tom turned to peer out and see who his visitor was, she said, "You don't know Jay Adams, Tom; it wouldn't make much difference whether I'd taken his cache or not, if he decided to go after me—or anyone else. He only travels one way—with plenty of experienced guns all around him."

She'd spoken in a dull monotone without looking up at him at all, so when he moved out into the front office, heaving the door closed on her, she wasn't immediately aware of it.

His caller was Otto Meuller, his coarse, strong features sagging with powerful but confused resentment. "I got to talk to you," said Meuller, dropping flat down upon a chair beside the desk. "This'll likely be the only chance I'll get to do it."

"Why?"

Meuller's little pale eyes lay upon McGrath without moving, like the eyes of a dead animal. "Because today I got a reason for comin' to town. The boys are over at the store loading a wagon with flour and other supplies." Meuller paused, then said, "They're watching me all the time, Tom. They're sly at it, but I'm no greenhorn; I've lived a long time with an eye in the back o' my head for skulkin' redskins. They're acting the same way. If I ride out, they slip along, keepin' an eye on me. If I send men out, they keep watch on them as well as on the house."

"Who, Otto?"

Meuller threw up his large hands. "I don't know who,

Tom; I only know who I think it is. But that's my secret for now. What I want you to do is—"

"Jay Adams, Otto?"

Meuller's words died. He studied McGrath awhile, then gave his head a heavy nod. "You're smarter than I ever figured," he muttered. "How'd you guess; the girl tell you?"

"No. Forget her for now."

Meuller's head reared back in a whip-saw motion. His little eyes glared. "That's exactly it," he bellowed. "That' s exactly what I want you to do, Tom. Set that blasted girl loose. Listen to me! I already told you—she can have that lousy horse. She can have ten more horses—and saddles with 'em—but you've got to set her loose. If you don't, that bunch is going to hang around until you do, and meanwhile they're going to hurt somebody. Maybe they'll kill me or someone else."

"Otto, that's why I can't turn the girl loose. They're here to kill her."

"Well, darn it all, McGrath, who's most important—the decent law-abidin' folks around Puma Station, or one little outlaw's girl friend? I tell you—turn her loose. You've got no call to hold her anyway."

"That's right; I haven't. At least up until now I haven't. I was counting on you to sign a complaint so I could do it legally, Otto, but you ran out of guts."

"*Guts!*" roared the bigger, older man, springing to his feet. "What are you talkin' about—guts! You confounded numbskull, those men are deadly. They're going to hit this town or kill somebody. I got two daughters out there with me. Suppose this madman decides he's goin' to start shootin' folks until you set his lady friend loose; suppose he starts, some kind of eye-for-an-eye business and shoots *women!* Tom, I'm tellin' you, unless you set that girl loose I'm goin' all over the range—and here in town too—tellin' folks what you're bringin' down on us."

Meuller went to the door, grasped the latch so hard his knuckles turned pale, and stood like oak, glaring down at Sheriff McGrath, who hadn't risen from in front of his desk.

"Otto, you listen to me for just one minute. Adams is here to kill Ruth Hall. There isn't a man in the Puma Station country who'd stand for that. But there's more; U.S. deputy marshals are on their way down here from Helena right now—today. I don't know when they'll arrive, Otto, but I'm going to wait. Then, if it's necessary, I'll hand the girl over to them and they'll take her away."

Meuller stood glowering for as long as it took to turn this over in his mind. The first indication Tom McGrath had that he might soften a little was when his hand on the door latch loosened slightly. "When'll they get here?" he growled. "Tom, I'm sittin' on a powder-keg out there. Those men could decide to hit my ranch any night. As it is, we're forting up after supper like the redskins were still around. It's hard on my daughters. It's even harder on me. I know what men like those, do when they take womenfolk captives. I'm not going to let it go that far."

Tom, pressed for a date, took a long chance and said the federal marshals should arrive within two days. He'd been tempted to say three days, but the adamant look on Meuller's face had made him, at the last moment, whittle off one day.

"Give it that long," he asked. "Otto, I don't want to see *anyone* get hurt. An' anyone also means Ruth Hall. Neither do you. You've got girls of your own."

"Hah! Not like *her,* McGrath."

"She's still a girl, Otto. You wouldn't want to see her dead carcass either—maybe with her face shot off."

Meuller's eyes dropped, then lifted again. "Two days, Tom. Then I'm goin' around and get enough folks to come in here with me to force you to get that girl out of Puma Station."

Meuller slammed the door so hard when he departed the tin cups suspended from nails over by the sheet-iron stove rattled and banged together. From beyond the cell-room door Ruth called out.

"Sheriff Tom . . . ?"

He swore fiercely under his breath, then went over and dragged back the oaken door to glower in at her. She was sitting on her bunk looking out at him, looking very small and very ragged.

"Thank you," she said softly, and showed him a little tired smile. "For all your meanness, Tom, you're a good man. I only knew one other as good in my life—and he's dead now. My father."

McGrath teetered on the brink of retreating, but in the end he stepped through, grabbed his nail keg and dropped down. "How the devil," he growled, "did you ever get yourself into such a jackpot, Ruth."

"Would you really like to know, Tom?"

"Well, for gosh sakes, haven't I been trying to sweat it out of you for nearly two weeks now?"

She smiled up at him again. "You know just about all of it anyway. I suppose, if you're struggling so hard to help, the least I can, do is give you some support. Isn't it?"

He looked dour. "It sure would surprise me if someone—just *anyone*—decided *I* wasn't the villain in all this."

"Tom, I heard what Meuller said out there just now. And he has reason to worry."

"His daughters are about all he's fretted over since—"

"No. That's not it," said Ruth, effectively silencing McGrath. "It's the money."

McGrath's mouth fell open as the implication hit him, hard. "Out—there . . . ?" he gasped.

She nodded. "They almost got me when they shot my horse instead of me. But I got away; found that trapper's camp and stole the clothes and the carbine. They tracked me, but where I grew up, Tom, we learned all the Indian tricks. I brushed out my tracks, hunted rocky places to walk over, kept in shadows and moved fastest at night."

"Lugging all that money?" He was astonished.

"It wasn't heavy. After all, it is money. Anyway, when I was just about at the end of my rope, I saw the Meuller ranch in the moonlight. I knew there'd be a horse for me down there."

"So you hid the money and went down to—"

She shook her head at him. "I got the horse first—then hid the money. I'll show you the place if you like."

He rocked back, and his nail-keg chair groaned ominously. "I'll be darned. Ruth—now tell me why?"

"It goes back three years, Tom, to the time I first saw Jay Adams. Have you ever seen a good picture of him?"

"No."

"He's handsome, Tom."

McGrath made a sniffing sound and sat more primly on his keg. She nodded at him, reiterating that the notorious outlaw was a handsome man. He said, "And you fell in love with him."

She studied McGrath's disapproving face a moment, then plunged ahead into her story. "My mother died when I was very small. My father was a crippled drover from Arizona. We had a section of cutover land and ran sixty cows. We fished and hunted and ate well, but we didn't have anything else. I grew up, Tom, and back in the hills one springtime I came onto this handsome, dark young cowboy in camp."

"Jay Adams?" he ventured.

"Yes. Only I didn't know him by that name at the time. I used to visit him secretly. Well, he went away, and I was very much in love with him. Then he came back a little more than a year ago. My father was sick that winter. He died in the spring. Jay helped me bury him. He moved into the ranch house. We fished and hunted and laughed a lot. Then I saw one of those newspapers with his picture in it. No coaches ever came near the ranch, so I'd never seen one before. He had it with him, and one night when he'd been drinking he showed it to me. He was bragging that night. It made me sick, all those killings, all that stolen money. I tried to get him to give it back, to go down to California with me where we'd start over." She looked away from Tom. "That wasn't very smart, was it? I'd never been in love before, Tom. I'd never known a man other than my father. . . . Well, to make a long story short, I decided to take the money—I knew where he'd cached it—and give it back in exchange for amnesty for him." She held out her hands, palms up. "And here I am—and he's out there somewhere with another wolfpack; only this time he won't be smiling. He'll kill me on sight. I know that now; for the first few days I was in here I refused to believe that, but now I believe it. I know him very well. He'll kill me on sight like he'd kill anyone else who'd do to him what I've done."

"It won't be that easy," muttered McGrath, feeling

embarrassed for her, and sad too. “They say Jay Adams is the smartest outlaw of them all, but I’ve never heard it said he’s fast with a gun.”

She clasped her hands together, not looking at him. He felt big and awkward, perching there on that rickety little nail keg. Their silence drew out long and deep before he rose and went back out into the office, closing the cell-room door after himself.

He felt tired. He also felt baffled enough to head up to the Angel’s Roost and take on a load. But instead he sat down and thoughtfully began cleaning his six-gun. That was a kind of therapy for a man in his trade.

The picture in his mind wasn’t a very pretty one, and for some reason he didn’t fathom, he despised Jay Adams personally. The fifty thousand dollars wasn’t as important as it had been earlier, but getting Adams was now twice as important without any consideration for the reward on his head.

Two days should do it, he told himself, if those cussed U.S. marshals would get here in that time. But that thought prompted consideration of another complication. They’d arrest Ruth Hall as an accessory after the fact; they’d have the legal right to do that on the grounds that she’d known for a year or more where Adams was hiding—where the money had been cached—and in all that time hadn’t made any attempt to contact peace officers.

He finished with the gun, leathered it, washed his hands and made a smoke. There was a way out of that for the girl, maybe. At any rate, although it wasn’t quite within his oath of office to think like this, he was doing it. He’d tell her to offer to lead them to the money in exchange for amnesty. They’d have a hard decision to make, but he was reasonably certain of their agreement.

Anyway, she'd been trying to get the money back when all this caught up with her in Puma Station. That would count in her favor, if she had to stand trial.

The trouble, of course, was that Jay Adams didn't mean for her to stand trial. Tom stood up, dropped on his hat and went out to get some dinner. He meant to bring her back a tray, too. It was odd, he thought, ambling up the plank walk towards the café—for once in his life having forgotten to lock the jailhouse front door when he departed—but here he was, sworn to uphold the law and having faithfully done so for ten years and better, trying to subvert it.

Well, maybe not exactly "subvert" it, but at least trying to keep it from bringing down its scales of justice against Ruth Hall.

When it was all over he'd probably have to take off the badge. He didn't dwell much on that; he'd been making a living before he'd ever put a badge on, and he'd do the same after he took it off.

At the restaurant he entered, nodded to some acquaintances along the counter eating, and shot the proprietor a look which the café man understood and at once moved into his curtained-off kitchen to implement. One of the townsmen eating in front of Tom turned and said, "There sure are some wild rumors floating around town." He winked and grinned broadly, indicating he didn't believe them at all. "For laughs you should've been up at the saloon this morning. They were saying old Meuller's organizing his riders like he was a general an' they was his soldiers."

Tom mechanically smiled as the townsman turned back to his meal and an indulgent little ripple of laughter passed back and forth among the other eaters at this preposterous idea.

CHAPTER 7

TOM DECIDED AT THE LAST MOMENT TO HAVE HIS supper at the jailhouse with Ruth and went into the kitchen to tell the café man that; tell him to double up on all portions and put aside two services. As far as the café man was concerned, it was still two suppers, and he charged accordingly. McGrath signed the receipt and departed, bearing his burdened metal tray. It never once occurred to him he might be walking into physical trouble, not even after he saw the jailhouse door was ajar, because he recalled not having locked it.

He shouldered his way inside, heard a sharp gasp and from the edge of his left eye saw something swinging toward him. McGrath acted instinctively. He heaved the tray forward and heaved himself backwards.

There were two of them, but the one who'd been watching for him at the door and had tried to brain him was so close that when the laden tray struck, he went down with a muffled squawk, putting himself fairly well out of it.

The second one, over beside the cell-room door with McGrath's big key ring, came around without actually going for his hand gun until he'd flung the key ring toward the door, perhaps in the belief it would distract McGrath. It might have except that he was distracted enough as it was, seeing those two men with their obvious intent inside his office.

He twisted away to avoid being struck by the key ring and went for his gun. The one on the floor, with soup and stew and lemon pie splashed over him, drew his gun, too, but McGrath kicked the man hard, dropping him over flat on his back. The other one had his weapon drawn but didn't fire. Tom had seconds to speculate about that. The man had no intention of firing; if he shot off that gun just once, the town would come boiling down to the jailhouse, and if he didn't get lynched for plugging their lawman, it would be a mighty miracle.

Tom, initially going for his weapon, straightened up and lifted his hand. He and the second man exchanged looks. The sudden turmoil was ended as astonishingly as it had begun. McGrath, by the roadside door, studied his enemy over by the cell-room door, then said, "Stranger, you kicked the hive this time. What do you think you're going to do with that gun?"

"Shoot you right between the horns if you give me any grief," growled the gunman, straightening up, loosening a little, beginning a careful, skeptical study of Tom McGrath. "Lift out that gun, Sheriff, and drop it."

Tom smiled mirthlessly. "Not on your life, mister. If you want this gun, you come take it."

The gunman dropped his glance, but the other one was as limp as a rag doll amid the sticky wreckage and ruin on the floor. Someone banged on the roadside door. Jim Astor's garrulous voice said, "Hey, Tom, you in there?"

Before his captor could order him to be quiet, McGrath answered loudly. "Sure. I'm here, Jim. Go get some men and surround the jailhouse. I've got a gun on me in here."

Astor's loud gasp was audible even through the

road—way door. Tom's captor said, "You louse. I ought to blow your guts up that wall!"

Tom said nothing, watching the angry man's thin, sharp features. There was an excellent chance that the gunman wouldn't do what he'd just said, but there was also an excellent chance that he would. What determined things like this, Tom knew, was the make-up of the man with the gun. Some were wildly unpredictable and would shoot even though they thought it very likely they'd afterwards be cut down themselves.

But what McGrath was gambling on was that this ferret-faced stranger wasn't that unbalanced variety of gunman. It was a mighty poor gamble, considering Tom had never seen the man before and could not know what type of psychopath he was facing. But the man's sharp, conniving features had prompted Tom to take his gamble.

They stood watching one another. After Jim Astor's departure, there wasn't a sound anywhere. The unconscious man remained out of it and unmoving. McGrath didn't recall exactly where his boot toe had landed; all he remembered was that he had put a lot of force into that desperate kick, and it had landed solidly.

"Come over here and open this door," snarled the gunman.

Tom raised an eyebrow. "Why, there's only a girl in there."

"I know who's in there, blast you. Quit stallin' an' do like I say."

It was too late; even the gunman must have realized that as a running herd of booted men came down the roadway. They might have been trying to be stealthy and quiet, but it sounded like a herd of wild boars lost in a muskeg swamp the way they grunted and bumped things out there.

Sheriff McGrath said, "Mister, put the gun down. They're thick as flies out there. They'll come busting through this door in a moment. If you're still standin' there with that gun in your hand, they're going to lead you down."

Out front, several men hooted back and forth, drawing in others as they approached the jailhouse front wall. McGrath told the gunman he had only seconds to make up his mind. The man on the floor groaned and rolled over, bringing up his knees and covering his stomach with his hands. Evidently that was where he'd caught McGrath's kick.

"It's not worth getting killed over," Tom said to the gunman. "Throw the gun down."

The man obeyed at the same time the roadside door burst inward and five men boiled in out of the night. They had cocked pistols and taut trigger fingers. Tom said, "Easy. Real easy." He dropped his gun.

Jim Astor, out front, looked at the groaning man, at the spilled food, at Tom McGrath, then back over at the tense unarmed man. "Tom, you forgot to lock the door?" he asked plaintively. It was such an innocuous question all the starch went out of the bad moment.

"Sure did," assented McGrath, going over to pick up the guns of the strangers. "Thanks, Jim. He wouldn't have shot me anyway, but this keeps me from having to kill him."

The range men put up their guns, looking narrowly at the upright prisoner. One of them said, "How you know he wouldn't have plugged you? He looks mean enough to me, Tom."

"He had no intention of shooting me," reiterated McGrath, looking his prisoner straight in the eye. "He didn't have his gun cocked."

Only two kinds of people aimed guns at folks and forgot to cock them; novices who were too rattled to realize that a single-action gun is useless uncocked, or men who knew perfectly well what they were doing and had no intention of firing. The stranger with the ferret-face was no novice. He wore a flesh-out holster, his gun was tied low, and his face showed the unmistakable signs of caution, shrewdness and calculating cold-bloodedness which spelled killer.

Other men, both range men and armed townsmen, pushed and shouldered their way inside out of the night. They almost filled the little jailhouse office. Several of them made growled, unflattering remarks about the prisoners. Two hoisted the groaning man to a bench and held him there, although he rocked back and forth hugging his middle and looking sick.

Jim Astor said, “You know ’em, Tom?” When McGrath shook his head, Jim said, “Well, it sure looks like they know you. Who’re they after?”

“The girl,” another armed man growled. “Who else?” This was a range rider, greying, granite-jawed, pale-eyed and rough. “Sheriff, any truth to the rumor Jay Adams is around somewhere tryin’ to get this girl out of your jailhouse?”

Tom went across to his desk with the key ring in his left hand. They all watched him. Obviously, Otto Meuller hadn’t held off after all. Tom turned and sat on the edge of his desk, eyeing the grey-faced gunman. “How about it?” he asked.

The prisoner said nothing. His eyes skipped from face to face. If he sought compassion he found none. A willowy, long-haired youth stepped over and nudged the man. “You deef or something? Didn’t you hear the sheriff ask you a question, mister?” The thin cowboy had a fist balled up.

The groaning prisoner raised a contorted face. In a husky whisper he said, “Adams is around. You set that girl loose or you’ll wish you had.”

It was a threat, but under the circumstances, coming in a half-whimper, half-whisper, it lacked all the hard menace a threat should have. The men looked at this one, acting more curious than fearful.

McGrath, satisfied the men had their answer, said, "That'll about do it, boys. Go on back up to Jim's place. I'll be along directly and stand the drinks."

They stamped out, talking among themselves, passing around tales they'd heard of the notorious Jay Adams. They were vastly impressed, McGrath saw that. Even Jim Astor, the saloon man, who had known his share of badmen, was impressed. When Tom closed the door on the last of them, he turned back with a wry expression.

"He'd better have better men than you fellers," he told his prisoners. "Otherwise he's likely to get buried here in Puma Station."

The slightly built one over on the bench was recovering from his injury and gradually straightening up. He gingerly felt his stomach region but watched McGrath. Most of the self-pity had by now been replaced with a venomous anger against the man who'd kicked him like that. He said, "It'll be a cold day in hell when a stinkin' place like this can stop Jay Adams. It's goin' to work out just the other way around, Sheriff. Jay ain't goin' to get buried here—you are."

Tom was hungry. He was also weary. "Stand up," he told the slightly built one. "Shuck your pockets inside out, toss down your hat and put everything into it. Then roll up your britches."

They both obeyed, and McGrath made a cigarette while he watched and waited. When they'd finished, he was satisfied they had no hide-out weapons and ordered them to step away from their hats and roll down their trouser legs. He lit up and asked where Adams was.

They snarled defiance at him without answering.

"All right. Where is their camp?"

They snarled again. This time McGrath exhaled a low gust of greyish smoke, studying the pair. "Once more," he said softly, straightening up as though he intended to move, "how many men are with Adams?"

The ferret-faced one curled back his lip, conditioned by now to defiance. Toni took two big strides and caught him alongside the cheekbone with a savage blow. The man went hard back against the cell-room door, then slid down it as though he were made of oil.

McGrath turned. The slightly built one raised a defensive hand, all the sneering defiance wiped off his face. "They's ten o' us," he panted. "Eight now that you got Tim an' me. Eight an' Jay hisself. An' he can get more any time he wants 'em."

"Where's the camp?"

"No one place, Sheriff; he keeps on the move. Sometimes we even move twice in the same day."

"What's he said about the girl?"

"He told us she stole his cache. He said he wanted her took alive, and when he got through sweatin' out of her what she done with his money, he'd let the rest of us do what we wanted with her. Then he said we'd have to shoot an' bury her." The man held out a timid palm. "I need a smoke, Sheriff, or a stiff drink."

McGrath handed the man his makings and turned as the other one moaned and flopped around, trying to make his head function and his limbs obey. The slight one asked for a match, handed back the tobacco sack, leaned for Tom to light his cigarette and shallowly inhaled as he ran a wise, sly look up and down Tom McGrath.

"He said he'd give us a thousand bucks each if we

could slip in an' fetch her out of your jailhouse. We agreed to try." For a second the renegades eyes lit up. "We darned near brung it off, too, only you come back sooner'n we figured."

Tom brushed all this aside. "What'll Adams do when you two don't come back?"

The outlaw didn't know. "He don't say much about stuff like that. But he'll know we been caught, an' that'll mean you fellers are more alert and knowin' than he figures you are."

"And . . . ?"

The outlaw took a deeper drag off his smoke. "I don't know. But you can sure as shootin' bet on one thing: he won't just sit out there somewhere cussin'."

"Who are the eight men with him?"

"Renegades," stated the prisoner candidly, wickedly smiling as though he enjoyed this. "Four or five are fellers who've been on the trail with him before. Me'n my pardner only picked it up along the trail when Jay Adams was lookin' for riders. We joined him over in Idaho. He promised us all plenty of loot."

"That ranch you've been spyin' on west of Puma Station: Why?"

"Why we been spyin' on it? I don't know. Like I already told you, he don't explain everything. But I got an idea."

"Keep talking."

"Well, that's where the girl stole a horse. We didn't see her do that, but we came through the night not more'n a half-hour after those range men discovered she'd took it. We sat up there in the trees an' darkness listenin' to the furor. Now I got an idea Jay figures the girl might've cached his money around that ranch somewhere, an' he's keepin' the place under watch just to make sure none of those range men find it and try sneakin' away."

Tom killed his smoke, scooped up the key ring, stepped to the side of the ferret-faced one and unceremoniously hauled the man to his feet with one hand. "Stand," he growled, propping the groggy man against the wall while he opened the cell-room door. "Now get inside," he commanded. They obeyed at once. He pushed them roughly on down the narrow hallway, saw Ruth watching intently, locked them into a cell and stopped outside her cage. "You heard?" he asked. She nodded. "I'll go get us another tray of grub," he said, and moved off. This time he didn't lock the cell-room door, but when he passed out into the darkness beyond, he turned and very carefully locked his roadside door.

The café man raised his brows but asked no questions. He went about making up another tray, and as before, Sheriff McGrath paid and walked out, heading towards the jailhouse. This time the café man went as far as his own doorway to watch. He scratched his head. That was a powerful lot of food for just two people to eat. He hadn't yet heard what had happened down there, but he would. In time the whole town would hear of it.

CHAPTER 8

TWO DAYS LATER, IN THE LAST RED GLOW OF EVENING, something occurred in which Otto Meuller would have been interested. Two solemn-faced, trail-weary strangers rode into town, put up their horses and shuffled on up to the café to eat supper. They didn't talk, not even to each other. One was a greying, black-eyed man named Hugh Tyre. The other was younger, thicker, more violent-appearing. His name was Frank Moore. Both were deputy U.S. marshals and they'd been a long while on the road. They looked short-tempered enough to tangle with a buzz-saw. The café man read them that way; after all, like Jim Astor and other merchants in Puma Station, they weren't young men any longer, and they'd been on the frontier many years and were still alive, which simply meant they could judge strangers correctly the first time.

While those two were eating an early supper, no one else came in until Tom McGrath appeared in the doorway. At first, neither Tyre nor Moore looked up, but when Tom called for the usual tray of grub, both men put down their forks, twisted and gravely ran a look from Tom's badge up and down. Then they went back to quietly eating.

McGrath studied their backs and the way they wore their guns. They didn't quite have the range rider look, yet on the other hand neither did they look, or act, like furtive men on the run. Tom had his reasons for being interested in strangers. He strolled over, stepped astraddle the counter bench and asked, "Which way did you boys ride in?"

Moore, the younger, beefier man, said, "North," and kept on eating.

"From where?" Tom asked.

Moore finally put down his fork and straightened back to look straight at McGrath. "From Helena," he said, and turned back the limp edge of his vest to show the star in a circlet pinned there.

Tom nodded and spoke his name. Moore and Tyre responded in a like manner, but none of them offered a hand to the others. Hugh Tyre leaned ahead to see around Frank Moore. "Any proof it's Jay Adams?" he asked point-blank. "We rode hard and fast to get here, Sheriff McGrath; sure wouldn't feel too happy if this turned into a wild-goose hunt."

The café man brought out Tom's tray. He stood up, took the tray and said, "When you're finished here come on down to the hoosegow." He walked out. Moore and Tyre turned to gaze after him, then gazed at one another and finally lifted their faces to the café man. He smiled a little at them.

"Tom McGrath's the best lawman Puma Station's ever had, an' if you fellers doubt it, go ask around town.

Hugh Tyre quirked up his thin-lipped mouth in a semblance of amiability. But his eyes didn't smile. "We sure don't doubt that, mister," he said soothingly to the café man. "We were just wondering about things is all."

The café man looked straight at Tyre. "Care for more coffee?"

Moore critically examined the café man. "Mister, is everybody as touchy about this sheriff as you are?"

The café man's smile died. He turned away, saying, "I'll fetch the pot and give you a couple of refills."

For Tom McGrath the encounter had been a surprise. Not that he hadn't been expecting federal lawmen, but he'd thought there might be a regular troop of them, not just two. And one of those—the man named Tyre—seemed almost past the age to tangle with someone with the reputation of Jay Adams. The other mean-looking, younger and heftier one looked more like a bandit than a lawman. There was an old saying that it took one to catch one. Frontier law, Tom knew, put more than just amusement into that old proverb; more than one badge-toting peace officer was a reformed outlaw. Some, like Jake Slade of Colorado and Wyoming, jumped back and forth between legality and illegality like a cat on a hot stove.

But there was no denying it, those two looked able to handle themselves in just about any environment. Tom took the food in, opened the cell door, closed it after himself and waited until Ruth had pulled up the small table. He already had his rickety nail keg in there. As he sat down he told her of his encounter in the café. She wasn't surprised, but she mentioned something about the speed they'd shown in getting to Puma Station. Tom had the answer to part of that.

"They load their horses into a cattle car on the train and go as far as the tracks'll take them, then finish up saddleback. It's still a long haul to Puma Station from the rail line, but it's a sight shorter than riding the full distance." He ate and thought, and eventually said, "Ruth, it's not my business to advise prisoners. In your case I'll make it my business. At the best estimate, Adams still has around fifty thousand of that railroad express-car money. They want it back right bad."

"I suppose so," she agreed.

"Then you make a trade with them." She looked up, violet eyes round and interested. "You offer them the money in return for amnesty from any charges they might want to file against you. No amnesty, tell 'em, no money."

From the doorway a man's thin voice drawled, "Sheriff, you sure run a cosy little jail here." Hugh Tyre was leaning there. Behind him Frank Moore was scanning the gun-rack, the desk, the benches, and finally turned to throw a flinty glance at Ruth Hall. He looked harder and opened his narrowed eyes slightly. Moore was surprised, which wasn't hard to understand; female prisoners weren't usually pretty or young or put together like a purebred Durham heifer.

Tyre kept looking at McGrath. "You give counselin' service too, I see. That's mighty obligin' of you, Mister McGrath. An' supper for two with your prisoners."

The other two prisoners were watching Tyre and Moore as though they'd just flushed out a den of rattlers. Moore shouldered past and went down to stand in front of the other cell, gazing in. "Who're these two?" he asked, without looking away.

"Riders with Jay Adams," said Tom, reached for his cup of coffee, emptied it and set the cup down again. "Mister Tyre, this is Ruth Hall. She's the one Adams is here to kill. She took his cache and tried getting away to hand it over to the law. They've been on her trail ever since, and they darn near got her a time or two."

"That's interesting," murmured the older man, settling his opaque eyes upon the girl. "But if that's so, Sheriff, why were you instructin' her to make a trade with us for amnesty from prosecution?"

McGrath stood up, opened the cell door and stepped through. He turned and very carefully locked the door; then he said, "I'll tell you why, Mister Tyre: because I believe her, an' she's guilty of no felony."

Moore strolled back to resume his interrupted study of the girl. She sat there watching them with anxious, troubled eyes. She had clearly catalogued Hugh Tyre as something she wasn't anxious to associate with. Moore she'd stabbed with one long, hard look, and afterwards ignored him. She evidently had arrived at her judgement of the younger, meaner-looking deputy marshal a lot quicker.

Tyre strolled out into the office, turned and waited for the other two. As McGrath locked the cell-room door, Tyre said, "Seeing you in there eating supper with her,

Sheriff, with the cell door unlocked, it looks a little odd that you're lockin' that other door now."

Tom felt his neck redden. He was beginning to take a violent dislike to Hugh Tyre, who was sarcastically cynical, and also to Frank Moore, who was just plain ornery. As he crossed to the desk, Tom said, "Listen, Deputies; let's have an understanding here and now. I don't like your forked-tongue talk, Tyre. And, Moore, if you want to be tough, right here and now'd be a fine time for us to see who's going to give the orders afterwards."

Moore stopped moving, studied Tom a moment, then rolled back his lips in a confident leer. "Suits me," he muttered.

McGrath withdrew his six-gun, put it atop the desk and was reaching for his hat to put it aside when Hugh Tyre, his voice no longer softly drawling, snapped at them both. "That'll be enough of that. After we've got Adams locked up in your jailhouse, McGrath, if you still want to test Frank, it'll be fine with me. Not before."

Tom was roiled and ready. He started forward as he said, "I told you, Tyre, *I* give the orders."

Moore was expecting trouble, but Hugh Tyre still meant to head it off without an actual showdown. He stepped swiftly between the other two and said, "Sheriff, it don't work, cow-county lawmen layin' down the law to federal officers."

"It works here," growled Tom, reaching to shove Tyre aside.

But the older marshal wasn't finished. "Frank," he barked, "keep away from him!"

Moore hesitated, not of a mind to obey, but he did, slowly and sullenly. He circled clear and halted over near the cell-room doorway. Tom turned on Tyre. "I've got two to whip and maybe I might as well start with you."

Tyre's right hand dropped down. "Don't try it," he warned. "Now listen, McGrath, for Gawd's sake. What's to be gained by behaving like—?"

"I told you, mister, you don't come into my town making off-color remarks about me or anyone else. And *I give the orders!*"

Hugh Tyre bobbed his lean head. "All right. You give the orders. Now let's settle down."

Tom stopped and dropped his hands. He and Frank Moore traded relentlessly sulphurous looks. Tyre said over his shoulder: "Frank, sit down." Moore sat. Tyre then reached up to thumb back his hat and look McGrath over more intently. "You're sure on the peck," he murmured. "Well, I reckon we did come in here sort of tired and raunchy. We apologize." Tyre smiled; it was a humorless smile with only his thin lips involved. Tyre's black, liquid eyes never seemed to smile, or for that matter to show any other emotion. Maybe they couldn't, being so black, or maybe he had long since had all the emotions wrung out of him. As far as Tom McGrath was concerned, it didn't matter.

Tom sat at his desk. "That girl goes free."

Tyre nodded, still measuring his man. "Don't see any other way, after you've advised her about how to do it. All right, Sheriff, the girl goes free—after she leads us to that cache."

Tyre went to a chair, sat, and crossed one leg over the other. Moore was making a smoke now, acting oblivious to the presence of the others. But when he lit up he slewed another sulphurous glare around at Tom McGrath.

"Where did she hide the loot," he asked, "and if Jay Adams is in the country, how come he hasn't found it yet?"

"Because I've held the girl in jail under protective custody, for one thing," retorted Tom. "And for another, because he wasn't able to spring her out to tell him where she hid it."

Moore turned laconic. "But she told you, eh, Sheriff?" The invidious way he said that made Tom's neck redden again. Hugh Tyre, recognizing the battle signs, growled at Moore to be quiet, to keep out of it, that he'd do all the talking. Moore subsided for the second time.

"She told me where it is," stated Tom, "but only in a general way, not specifically. In other words, I can't take you out there in the morning and point out the cache."

Tyre nodded, showing patience one wouldn't have thought he'd have possessed. "But the question is: Will she lead us to it? I've had them do that before, Sheriff, and at the last moment have a change of heart."

"She won't *lead* us to it, Marshal, because I won't let her out of the cell she's in right now. The second she rode out with us tomorrow, Jay Adams's spies would see her. They'd pick her off sure before they'd let her lead's to that cache. But I'll ask her to draw me a map. When the three of us'll go get the money. Anything wrong with that?"

Tyre smiled that thin, bloodless smile again, and wagged his head. "Nothing at all wrong with it, Sheriff McGrath. Our orders are to get the money back."

"That's all?"

"No. We were empowered to barter a little, if we had to, in order to recover. All right, we've agreed to barter her freedom for the loot. Then we were specifically told to hunt down Jay Adams and fetch him in dead or alive, but under no circumstances to return without proof that we'd met him."

Tom spoke flatly. "Marshal, there are eight other renegades riding with Adams. Those two men in the other cell told me he'd hire more men if he felt he needed them. Those are pretty stiff odds against three lawmen."

"You mean you can't make up a posse hereabouts?" asked Hugh Tyre, holding the cigarette away from his face as he awaited Tom's reply.

"I can get up a posse," exclaimed McGrath. "But I'm not leaving the town—or this jailhouse—unguarded. And I'm not going after that loot until I'm darned well satisfied about your credentials."

Frank Moore stiffened, on the verge of retorting, but once more Hugh Tyre demonstrated who was in charge. He laid a cold, withering black stare upon Moore, and the younger officer sat and sizzled, but said not a word.

Hugh Tyre's patience, however, was very thin as he dug out some papers, carefully unfolded them and laid them on the desk. "Pictures and letters of introduction," he said coldly. "That little card there is the date of appointment, when and where. Satisfied, Sheriff McGrath?"

Tom read the documents slowly and purposefully before handing them back with a little nod. "I'm satisfied. Now if you men need some rest, there's a rooming-house up the road. There's also a bath-house out back of the rooming-house. If you want, I'll be waiting for you with my horse saddled right after breakfast in the morning. I'll have the map of where that loot is buried; we can go get it."

Tom stood up. Hugh Tyre, refolding his papers, also rose. Frank Moore strolled over from the far side of the room, reached for the door and said, "An' if she won't tell you where the money is, what then, Sheriff McGrath?"

Tom smiled. "Dream on it, gentlemen. I'll be waiting out front in the morning."

After the federal deputies had departed he went into the cell room with a paper and pencil. Ruth shook her head at him. "I don't have to draw a map, Tom, and you wont need a map to find it. I threw the saddlebags atop the barn roof where no one would think to look, unless it rained hard and the roof leaked."

Tom was flabbergasted. So were the pair of intently listening outlaws farther down in another cell.

CHAPTER 9

TOM WAS A MAN WHO TOOK AN OATH SOLEMNLY. HE was waiting with his horse saddled outside the jailhouse an hour or so past sunup. The deputy marshals saw him there and, without waving or calling over, turned and went to the livery barn after their own animals. A lounging silhouette in the doorway behind McGrath made a pithy observation.

"Like you said, Tom, they're a real pair of beauties. It must hurt 'em to smile."

McGrath watched the lawmen disappear into the livery barn, then turned. "Jack, you just make darned sure of what I told you. No one comes inside while I'm gone. Adams is smart. He's made dunces out of a lot of people. If he gets in after we're gone, he won't just shoot Ruth; he'll also take care of you."

Kirk didn't look particularly concerned. "No one comes in; that's the size of it. But you better get back before the lady and I starve to death. By the way, does she play poker?"

Tom saw the federal men emerging from the livery barn astride, and swung up to ride forth into the center of the road. He turned while Hugh Tyre and Moore were still thirty feet away and led off, taking them out of Puma Station by the north roadway. After that he booted out his beast and loped west and just slightly south.

Tyre and Moore caught up, one on each side, when they were a couple of miles out, and McGrath slowed to an unpleasant trot. Just for meanness he kept at that gait. Western horsemen were not fond of trotting. They either loped or they walked; sometimes they'd gait their horses to shuffle along in a fast little running walk called everything from a "sopentater" gait to the "Tennessee walk." But trotting was not a popular gait at any time.

Eventually Frank Moore growled at McGrath. "Say, if we're in a hurry let's gallop. If we're not, let's walk."

Tom slowed. The land was warming up. Heat would come shortly, and to men conditioned by Montana's long, bitter winters and chilly springtimes, heat caused suffering. They saw trees and brush, broken erosion gashes streaming like wounds across the scarred belly of the plain, and they also started a little bunch of pronghorns, which normally they couldn't have gotten within rifle shot of. But these fleet critters, browsing in one of the erosion gulches, didn't see or even hear the riders until their shadows passed overhead. Then, with a shrill whistle, the biggest buck antelope dived out there and set the pace, his flag up, his white rear end moving with the powerful rhythm of the fastest four-footed animal on the North American continent.

"There goes supper," remarked Hugh Tyre, watching the animals bound away at something like seventy miles an hour. "When I was a kid, one time," he mused, "I caught a fawn and raised it around the homestead. Sure made a good pet."

McGrath turned, scanning the cruel, uncompromising face with its oily obsidian eyes. This wasn't the kind of talk he'd expected. Moore saw McGrath's expression and said, "He's human, Sheriff. Maybe he don't look it, or always act it, but he's human."

Marshal Tyre looked at them both, then changed the subject by pointing to the rough, rugged mountainscape to the north. It seemed to make a huge curve from the east around the west. "If Adams is in this country, that's where he'll be." Tyre dropped his arm. "Did you ever pursue a man so long, Sheriff, you got to know where he'd head for and what he'd have for supper? Well, Jay Adams has been our special target for over a year now. He's a man who likes the mountains. The rougher they are, the wilder and more remote, the surer he is to be in there somewhere."

Tom recalled what Ruth had said. She'd been in the mountains of her Snake River country when she'd stumbled onto Adams' camp. He also recalled that when Adams had gone to earth after the Nebraska train robbery, he'd headed for those same rough mountains again.

As they rode toward the Meuller place, Tom speculated. Actually, he knew nothing about Jay Adams, the man. He knew probably as much as the next man about Jay Adams the outlaw, but all that had required had been an ability to read. Newspapers for the past five years had considered Jay Adams the best news since the demise of the James boys, and the killing and imprisonment of men like the Youngers, Sam Bass, Slade, Billy Thompson, the English-born Texas gunfighter, and his more notorious brother, Ben Thompson.

He was so lost in thought that when Hugh Tyre asked whose ranch buildings those were on ahead, he failed to hear until the question was repeated. "Otto Meuller's place," he answered. "That's our destination."

Frank Moore grew curious. "Is that where she hid the loot?"

Tom said, "Yes," to Moore, then turned to Hugh Tyre. "If Adams likes mountains so well, maybe that can work against him around Puma Station. What other stuff do you know about him that could help us track him down?"

Tyre looked coldly amused. "Well, he's a dead shot with either hand, and is so fast with a gun folks say you can't even see him draw."

Tom turned this over, too; for some reason he couldn't now recall, he'd had the impression Adams wasn't fast. "What else?" he asked.

Tyre's cold grin broadened. "You're warmin' to the job ahead," he murmured with approval. "Well, he's a devil with the ladies. We tried to use that against him once, over in Wyoming. He left the lady waitin' in her buggy, and the next thing we heard he was in Colorado. I reckon he doesn't really *like* 'em; he just uses 'em."

Frank Moore said, with apparent meaning, "He sure knows how to pick them, though."

McGrath studied the Meuller place as they rode in closer to it. He was beginning to have an unorthodox idea. He knew something neither Tyre nor Moore could know: Otto Meuller's daughters were very close to their twenties, and as far as Tom McGrath could recall, he'd never once seen them at a social function in town unless their father was along. To girls like that, a man as handsome and fascinating as Jay Adams could make all the difference in the world.

Conversely, to Jay Adams, those two love-starved females could be very useful. What Tom McGrath wished to determine now, when he talked to the Meullers, was elemental: Had either—or both—of Otto's buxom daughters been shadowed and charmed by Jay Adams, who would certainly not pass up the opportunity, if it existed, to make friends with someone from the ranch where he thought it highly possible Ruth Hall had hidden his stolen money.

Tom began whistling. For some reason he didn't bother even to try explaining to himself, he began feeling much better about all this. He kept on whistling right into the ranch yard of the Meuller place, and for that reason they were met down by the barn by Otto himself, and two interested cowboys who'd evidently been working inside.

McGrath saw the bitterly defiant glare Meuller bent upon him and totally ignored it to say, "Otto, this is Deputy U.S. Marshal Hugh Tyre. This other one is Deputy U.S. Marshal Frank Moore."

Meuller and the lawmen took one another's measure while Otto failed to move forward to offer a hand and neither of the deputy marshals offered to dismount for that purpose.

McGrath ran a slow, sustained look far out, up along the forested ridges and lower down, where there was less wood and more grass or underbrush. It was likely that someone was lying up there watching. Tom hoped there was. He asked Meuller how many of his riders were on the home place.

Otto pointed. "Those two men over in front of the barn. The three of us have been out back repairing some corral stringers. My other three men are on the range."

"Are your daughters in the house?" asked McGrath, turning next to examine the large old low-roofed ranch house southward across the yard from the barn area.

"One is," stated Otto. "The other is out riding. McGrath, what is this all about?"

Tom looked down at Meuller, didn't say a word and reined over to the lowest eave of the barn, which happened to be at the northeast corner. While the others watched, McGrath stood up on the seat of his saddle, eased over and with a powerful jump caught hold of a rafter-end which he used to swing himself atop the barn.

Meuller and his pair of range riders stood with their mouths open. Otto sputtered as though he'd growl at Tom McGrath, but he didn't; he simply stood there watching the sheriff stalk across the sloping plane of his barn roof, wearing an expression which clearly showed he thought Tom McGrath's mind had snapped under the recent pressures and tribulations.

Tyre and Moore showed less bafflement and more hard, practical interest. When McGrath sank from sight up there they could still hear him walking. It was a huge old barn; the roof area to be covered was considerable, with many hips and valleys.

Tom returned around the westerly side, stalked back down to where he'd gotten up there, bent to consider the height, and flung down the set of saddlebags he was carrying, following them down by suspending himself from an eave until his feet touched a topmost corral stringer.

The men in the yard hadn't moved. Hugh Tyre and Frank Moore seemed poised to do so, but they didn't.

Otto Meuller drew forth an immense blue bandanna and mopped his forehead and neck with it. There was heat now, but not too much, actually. At least it didn't appear to bother any of the other men; or if it did, they were concentrating so hard on Sheriff McGrath as he came toward them with those old saddlebags across his shoulder, they didn't feel any heat.

Hugh Tyre spoke first, shooting a dubious glance roofward. "I suppose it was a good place, but darned if I'd have the guts to cache anything in plain sight."

"Of birds?" asked Tom, halting beside his horse and looking across at the house,. "Otto, try invitin' us in for a glass of cold water."

Meuller looked doubtful. "What are you doing?" he mumbled. "Who put those saddlebags atop my barn, and why did they do such a thing?"

"They were in a hurry," explained McGrath. "Come on, deputies; on the porch over yonder." He led the way across Meuller's large well-kept ranch yard and looped his reins at the rack out front, went up under the patio roof, took the saddlebags off his shoulder and, while awaiting the others, hefted them.

Ruth had been correct; they weren't heavy at all, almost as light as though they were empty. McGrath's heart gave a little lurch in its dark place. If they were empty, Ruth and Tom McGrath, were in trouble up to their ears. He had to fight down an urge to unbuckle each bag and swiftly peek in.

Otto said something to his pair of riders. They retreated back toward the barn, leaving Meuller, Tyre and Moore to come up onto the patio where Sheriff McGrath was waiting. When they were all up there, Tom said, "Otto, seen any watchers the last day or two?"

Meuller shook his head. He was a man of just one capability at a time; right now he was interested in those saddlebags from the roof of his barn to the exclusion of everything else. "Haven't seen any," he muttered. "Even my men haven't seen any on the range. I don't understand it at all, but it suits me fine if they've gone away."

Tom McGrath hadn't asked that question idly. He already knew, because Meuller had told him, that one of his daughters was riding alone on the range. He'd asked the question to substantiate a suspicion he'd been forming since entering the Meuller yard: That there wouldn't have to be any watchers up there now because Jay Adams had a spy right in the Meuller household—one of Otto's love-starved daughters!

Hugh Tyre frowned. "Open it," he said. "Daydream some other time, Sheriff McGrath."

Tom stepped back to a little weather-checked patio table, flung down the saddlebags and methodically unbuckled each side; then he tilted the things slightly to let their contents trickle out.

In a whisper of crisp sound the table-top turned green. Bills of large denominations slid forth. Some of the money still had the mint-wrappers around little bundles.

No one spoke for a while. Tom McGrath let them have a good look, then began stuffing the currency back into the saddlebags. Finally Otto whispered in a shocked voice: "On top of my barn . . . ? What kind of a hiding place is that for a fortune?"

"Good enough, I reckon," stated McGrath, looking at Tyre and Moore. "At least it wasn't a very likely place for anyone to look, apparently, or it'd have been gone before we got to it."

Hugh Tyre reset the hat atop his head and looked

across the table at the others. He was obviously balancing upon the knife edge of some decision. Frank Moore, too, seemed changed a little by those saddlebags.

Frank said, "Hugh, that's one half of it," and McGrath mistakenly thought Moore had been speaking about the money as being one half of what had been stolen. But Tyre's answering remark let Tom know precisely what the two meant by their words and expressions.

Tyre said, "Yep; we've now collected one half of a bad debt. Jay Adams is the other half. Let's get this money put safely away and go after the other half."

Meuller pointed. "Is that money Adams'?"

The three lawmen all nodded at Otto, and the cowman dropped his arm, looking from one of them to the other. "Now I'm beginning to understand. He knew it was around here some place; *that* was why he kept spies watching all the time. It wasn't to raid the ranch; it was to get back his money." Meuller was relieved. "All right; now you take it away from here, Tom. Take it all the way into Puma Station and cache it there. Don't you dare hide it on my ranch—anywhere."

McGrath agreed to do just that, and went walking down across the yard with the saddlebags loosely slung across his shoulder as he headed for the barn tie-rack where his horse patiently waited.

Tyre and Moore also went down there and got astride. Just before the three of them turned to head out, McGrath said, "Otto, do me a favor. Come into town this afternoon. I've got something I'd like to show you."

Meuller procrastinated, darkly scowling. "I've got work to do, Tom. A place like this doesn't run itself."

"Come in anyway, Otto. What I've got to show you

will be a real eye-opener. It's pretty serious and concerns you too."

Otto would have asked the obvious question, but McGrath turned his horse and loped away.

CHAPTER 10

THEY RODE BACK TO TOWN LIKE FUGITIVES, ACUTELY conscious that they could very well be under constant surveillance. Tyre said it puzzled him why Adams hadn't just gone down to the Meuller place with his outlaw brood and sacked the place, hunting his money. Tom McGrath's answer to that was brief.

"He still wouldn't have found it. Who'd ever climb atop a barn looking for hidden money?"

Frank Moore sided with McGrath, but he also said that if Adams had had a spy up there today, he'd undoubtedly have spotted McGrath bringing those saddlebags down off the roof. "And it sure wouldn't take Adams long to figure out what was in those bags, either."

Hugh Tyre twisted in the saddle, sweeping the countryside with an intent, searching look. He didn't say it, but he was nervous.

Nothing happened, though. They reached town a little after high noon and put up their horses before going to the jailhouse. Jack Kirk opened the door for them with a cocked six-gun in his fist that made Tyre and Moore look disapprovingly at the Two-Bar cowboy. They'd had no idea Sheriff McGrath had left anyone guarding his prisoners.

Tom introduced the three men to each other, then closed and barred his roadside door, up-ended the saddlebags atop his desk and heard Jack Kirk gasp. They took their time counting the money. In fact, when it turned out to be seventy-five thousand, which was twenty-five thousand more than they'd expected to find, they made a slower and more methodical recount. After that Hugh Tyre said he thought Adams had either included some other cache with this money, perhaps from another robbery somewhere, or else he thought it likely Adams had held out on his gang when he'd paid them off.

Tom wasn't overly concerned with the amount of money. He was, he told them, more concerned about where to hide it all. There was one safe in Puma Station —at the stage office—but it was small and old and been pried open twice in recent years by robbers.

"That's out," stated Tyre with strong emphasis. "It's not going to take him long to find out we have his money. He'll certainly make a try for it. If some two-bit renegades pried the stage company safe open, think what a seasoned dynamiter like Jay Adams could do."

"I have thought about that," exclaimed Tom McGrath, "and it's not the money that worries me. When he blows a safe he has a bad habit of overcharging his loads. People get killed, and there are bad fires. None of this is going to happen in Puma Station."

Tyre nodded agreement. "Then where'll we hide it?"

Tom sat down and furrowed his brow in thought. Jack Kirk, still somewhat baffled by all this, asked Tyre and Moore for particulars. They related where the saddlebags had been and a few other of the details surrounding them. When they were finished, Jack said, "Hey, Tom, take this money out of town an' tell the fellers at the stage company to leave their safe door open so if Adams comes, he won't have to blow it and maybe burn down the town."

McGrath considered the Two-Bar cowboy quietly. He had already made these identical plans. It was where to conceal the money he hadn't yet worked out. "Where," he asked, "outside of town?"

Kirk was nonchalant about that. "Any number of places. That's no problem." He saw the three sets of eyes on him. "Under the altar in the church down there below town to the south," he said. "Or send it out with the local mail when the stage leaves this afternoon. Simply mail it to yourself care of General Delivery to one of the neighboring towns."

Frank Moore was beginning to smile upon Jack Kirk. Even Hugh Tyre looked differently upon the youthful range rider. Tyre said, "The church altar, Sheriff. There isn't a less likely place on earth for a man like Jay Adams to go hunting his cache. And it'll still be where we can lay our hands on it in a hurry." Tyre's face suddenly got crafty. "Only none of us can walk down there in broad daylight with those saddlebags; he can have spies right here in town."

"Wait until after nightfall," said Jack Kirk, leaning upon the front wall, making a smoke. "One of us can slip down there with it after dark, and the other three can slip around to give him plenty of forty-five calibre protection."

No one said so aloud, but the others appeared to agree with this idea. Frank Moore then said he'd gone about as long as he intended to go without food, and this of course brought up the matter of guarding the money until dark. It was decided that Moore and Tyre should go out and eat, and after they returned Tom McGrath and Jack Kirk would do the same. In this fashion the money would always be adequately guarded.

After the deputy U.S. marshals had departed, Jack made a dry remark about them. "Never met a pair quite like those two before, Tom. Seems like they want folks to dislike 'em."

McGrath wasn't concerned and shrugged off Tyre and Moore. "Meuller's coming in," he said, gazing over at the cell-room door. "I'm banking he'll bring his girls."

Kirk eyed McGrath without comment. He slowly followed the line of McGrath's stare. The soft tone of voice that astonished Jack Kirk.

"I'm going to arrest one of his daughters—I think."

"*You think?* You been out in the sun too much lately, Tom. Otto'll take you apart thread by thread if you so much as just look at his girls, let alone arrest . . . What for; what did she do?"

"Nothing. At least nothing the law's concerned with. But Otto'll want her alive and unharmed, so we'll just call it—"

"Yeah, I know," grumbled the Two-Bar cowboy. "We'll just call it protective custody. Say, you're goin' to have a regular harem if you don't quit usin' that excuse for arresting folks." Kirk drew up a chair, sat down and said, "Out with it; what's eatin' you?"

"I've been piecing together some stray facts about Jay Adams," replied McGrath. "He knew that money was somewhere close by, possibly at the Meuller place, and it'd be a lot more sensible to have someone inside the place, on his side, than to waste a lot of time spying and hoping."

Jack's face suddenly brightened. "One of the girls," he exploded. "Why didn't I think of that?"

"You just did," responded the sheriff, who hadn't been allowed to complete his statement. He arose, passed over and unlocked the cell-room door. Instantly, all three prisoners looked out at him. He ignored the two men and stepped up to unlock Ruth Hall's cell; then, without a word, he beckoned for her to follow him into the outer office. She complied, looking anxious but—uniquely for a woman—she didn't start spouting questions.

Tom relocked the cell-room door, pocketed the key and pointed to a chair for the girl. She dutifully sat down, looking quickly from one of them to the other. Tom took his time; he sat down, made a smoke, lit it and heaved a mighty sigh. Then he said, "Did you know how much money was in that cache when you took it?" She shook her head at him.

"Just an awful lot of it, and all in large bills," she answered. "I didn't have time to count it anyway."

Tom said almost reverently, "Seventy-five thousand dollars."

His words fell into the silence like little steel balls. She gazed big-eyed at Jack, at Tom, then leaned her shoulders against the back of the chair. "That's an awful lot of money," she murmured, looking awed.

"But it's not worth dying for," said the sheriff. "Ruth, I think there's another girl mixed up with Jay Adams. She'll be coming into town directly. Now then, I'm going to arrest her. There'll likely be fireworks when I do that. Her pappy's a big, rough bear of a man. All I want you to do is try and keep this girl from crying her eyes out when she's put in the cell with you. I'd take it kindly if you'd do whatever you can to sort of make her feel—"

The roadside door opened. Tyre and Moore walked in, blinked at Ruth, at the guns in the hands of McGrath and Jack Kirk, then came on in and softly closed the door. Hugh Tyre, glancing sidelong at the girl, said,

"Sheriff, it was right thoughtful of you to have her waitin' when we returned."

Tom's neck reddened. "I told you, Tyre," he said softly, standing up and leathering his gun, "I didn't like your sarcasm."

The older man raised a hand. "Wait a minute," he said swiftly. "I apologize. I just got that habit is all." He squinted at Ruth. "You bought your freedom, miss, and you—or someone—sure paid a high price for it. Did Sheriff McGrath tell you how much was in those saddle bags?"

Ruth nodded, ignoring Frank Moore, who was looking straight at her in his rough and hungry way. McGrath dropped his little bombshell by telling Tyre and Moore what he proposed to do the minute Otto Meuller walked in. They were surprised; they were also baffled, so McGrath explained all over again, for their benefit, what he thought might be taking place. Then, finally, Hugh Tyre's expression changed. He even looked pleased, in an acid way, and said, "Sheriff, I got to upgrade my opinion of you for the second time today."

McGrath caught Moore looking at him with a slight show of rough amusement, and turned away to put out his cigarette. He was turning back to say something when a deep, rumbling voice out in the roadway profanely told a team to halt.

Jack Kirk nodded gently in McGrath's direction. From this both Moore and Hugh Tyre got the impression that Meuller had arrived. Tom dug out the cell-door key and tossed it to Kirk; he then jerked his head cellward to Ruth, who immediately rose and went over to be locked in again. At the door she looked back. "I hope you're right," she said. "There's no one on earth who can tell that girl with better authority what a fool she's being."

Otto stamped into the office, looked slightly surprised to see all three lawmen standing there evidently awaiting him, and when Jack returned to lock the cell-room door Otto said, "What's the matter; not enough work out at Two-Bar to keep you from loafing in town?"

Kirk enigmatically smiled, said nothing, and watched McGrath step to the door behind Meuller, look out at the parked top buggy, then jerk his head. Jack strolled on over by the door. Frank Moore, guessing what McGrath was thinking and having already taken Meuller's measure, also walked over to block the exit.

Tom said, "I see you brought the girls to town with you, Otto."

Meuller, sensing something here, looked darkly at the array of muscle in front of the roadway door. "I always fetch them along, but now with that scum skulking around, I watch them closer than ever."

"Not close enough," McGrath said. "Which one was out riding this morning?"

"Helen. She's been going riding every day for the past week or so. What of it?"

"You weren't keeping very good watch over her then, Otto."

Meuller's little pale eyes slowly turned raw and hostile. His big fists clenched at his sides. "McGrath, what are you hinting at? If it's what I think, I'll break every bone in your body."

"Make it *three* bodies," snapped Frank Moore, who evidently enjoyed this kind of fight.

"You shut up and keep out of things," snarled Meuller, glaring savagely. "Tom, you say right out whatever's on your mind."

McGrath paused, took in a big breath, looked left and

right, then said, "Otto, I'm going to arrest Helen and hold her in protective custody."

"You're—*what!*" Meuller's words made the windows quiver.

"Ask her to come in here, Jack." As Kirk turned to obey, Otto roared again, telling Jack if he opened that door Otto would break his neck. Jack pondered, one hand on the door latch, the other hand lying lightly on his six-gun.

"Man, I'm only a deputy here, a sort of part-time deputy at that, and that makes me about as high as a swamper over at Astor's saloon. Besides, if you did that I wouldn't be able to button my collar. Mister Meuller, I got my orders." Kirk opened the door.

Otto let out a bawl like a charging bull and lunged. Jack prudently didn't turn back; instead he ducked on outside and closed the door on Meuller. Frank Moore grinned from ear to ear and swung a solid blow that crunched into the bone of Otto Meuller's head below the ear. That hard a blow in that particular area should have folded the cowman up. But all Otto did was blink, turn with another roar of rage, and hurl himself straight at Frank Moore.

Hugh Tyre had his hand on his holstered gun across the room. Tom McGrath jumped clear as Meuller lunged blindly past him, and rolled his shoulders in behind the blasting right that caught Meuller in the same place, under the ear. Only this time Otto's legs turned rubbery. He put out a pawing hand to brace himself, and caromed off the wall as much from the power of that strike as from his forward momentum.

Frank Moore, back-pedaling fast, hooked both spurs in the lower rung of a chair and crashed over the piece of furniture backwards, still making futile movements to get clear.

Otto staggered around. McGrath said, "Otto, cut it out. Let me tell you why this is necessary."

But Meuller shook like a bear to clear his head, drew back his lips in a hideous grimace and took a step toward McGrath. This time Tom hung one on the point of Meuller's uncovered jaw, and the big cowman fell. When he hit the floor Frank Moore kicked clear of the broken chair and sprang up, staring in awe at the downed man.

"Hugh," he gasped, "I hit him with everything I had."

Tyre said nothing; he too was fascinated by the incredible durability of Meuller. But Tom McGrath looked over at Moore and said, "It wasn't half enough, Moore. You remember that next time we have a free moment to finish that discussion we started yesterday when you started walking roughshod over me."

Jack Kirk poked his head in, rolled his eyes downward, then said, "Hey, Tom, you better tie that old grizzly bear. On second thought he'd chew through ordinary lass rope; you better chain him."

McGrath flexed his knuckles. They were sore but seemed otherwise unimpaired. He said, "Fetch Helen in here, Jack. In fact, fetch both the girls in here."

Kirk nodded and ducked back out through the door.

CHAPTER 11

HELEN MEULLER WAS THE ELDER OF OTTO'S daughters. The other girl, Beulah, was buxom too, but she lacked the blonde sparkle of Helen. But when they walked in and saw their father being hoisted into a chair by Tyre and Moore, neither looked very vivacious.

Helen rushed ahead, bent down and dabbed at Otto's slack mouth where a little trickle of burgundy ran. Beulah stood beside her battered parent, looking astounded. Probably this was the first time either of them had ever seen their father bested.

Tom explained that Otto hadn't wanted the girls brought into the Jay Adams affair, and watched Helen's face when he said it. The reaction was so obvious he didn't have to be watching to see it. Still, big, buxom, blonde Helen Meuller didn't turn on McGrath until her father put up a big hand to probe his chin and jaw numbly. Then she straightened up, looking fire at Tom. He didn't give her a chance; he said, "Helen, you're under arrest."

The big girl's eyes widened in pure astonishment. "Under . . . Tom McGrath, you must be out of your mind. When Paw comes round and you tell him that, he'll—"

"No'm," stated Hugh Tyre, eying the big girl with strong appreciation for her merits. "If he starts it over again, next time he's goin' to get pistol-whipped."

Beulah had her hands to her throat. "Tom," she whispered. "Why?"

McGrath nodded at Helen. "Tell Beulah," he exclaimed. "Tell her why you've been riding in the hills lately, Helen."

It was of course a shot in the dark, but the watching men saw the color leave Helen's face. They also saw her hands tremble when she attempted to turn her back on McGrath, facing her father. He reached, turned her back gently by one shoulder and said, "That's why I'm arresting you, Helen. Not for being any part of his gang, but for your own safety; we call it protective custody."

Helen said, "Tom, what good will that do; I'll get out

some day, and I'll find him." She turned fully, defiantly, toward McGrath. "I've waited almost too long as it is. Paw'd never allow us to look at a man. And he's the most gracious, handsome man I've ever seen."

Tom saw Otto's eyes open, rivet themselves upon his daughter's back; at the same time Otto's huge fists clenched until the knuckles were white. To forestall an outburst, McGrath said, "Helen, how did you meet him?"

"On our range. He was riding alone. He just happened to be going my way."

"I'll bet," muttered Jack Kirk, then subsided under Helen's withering stare.

"He didn't lie to me, Tom. He told me right off his name was Jay Adams."

"You should've turned right around," said Hugh Tyre dryly, "and run as hard as you could for home."

She didn't even throw a withering look at Tyre; she simply said, "Tom, he's not at all what you think. He's so kind, so soft-spoken and gentle."

Otto's big fists were lying slack now. His eyes appeared steeped in some private misery as he considered his daughter. His face turned grey and slack. The other sister, Beulah, moved in to lay a hand lightly upon old Otto's shoulder. He didn't seem even to realize the hand was there, or Beulah either, for that matter. He was looking and listening in a stricken manner.

"Helen," said Tom McGrath, "I have another girl locked up in here. She also knew Jay Adams. I'm going to put you into the same cell with her. I want you to listen to what she can tell you."

"It won't change me, Tom. I never knew a man like Jay existed. I thought all men were like—well—Jack Kirk there, and the other range men."

Old Otto said, "Helen . . ." and heaved himself up out of his chair. "Helen, you should listen to me now more than ever."

She turned on him like a lioness. "I've listened to you until I'm an old maid, Paw. It'll happen to Beulah, too. Well, I'm not going to listen to you any more—ever. Paw, I *love* him. I . . . I'd have helped him get his money back, too, if I'd known where that vixen hid it. He told me about her; how she tried to sneak away and turn him in for the reward. How she tried to poison him, and once she even tried to shoot him."

Tom McGrath got off his desk and shuffled over to unlock the cell-room door again. When he swung the panel wide he said, "Otto, if you'd listened a few minutes ago it would have helped a lot. Now get away from her; she's goin' into a cell."

Otto didn't even look at McGrath. The stricken expression was still across his craggy features. "Honey, I was only tryin' to protect you'n Beulah. I wanted just the right boys, you see; boys who'd make decent homes and—"

"Paw, I never wanted a *decent boy.* Maybe when I was thirteen or fourteen. But, Paw, I'm twenty now."

"Not for another two and a half months," exclaimed Otto stubbornly.

"I'm *twenty,* Paw, and the kind of man I want is out there trying to recover some money that's been stolen from him. And I'll go with him anywhere at all!"

Tom McGrath moved in. "Helen, through that door and straight ahead."

The girl turned, saw Ruth Hall watching her from inside the cell, and stopped. "I won't be put in with that woman," she exclaimed.

Tom had another empty cell, the one separating Ruth

from Adams' two renegades. But he didn't make any move to incarcerate Helen Meuller there. He simply pointed toward the cell containing Ruth and said: "March!" From the corner of his eye Tom saw Otto take a step after his daughter and turned toward the older man. Otto stopped, dropped his arms to his sides and sagged. Beulah came over to him. As before, he seemed scarcely conscious of the other girl's presence.

Tom locked Helen in and returned to the outer office where he carefully locked the second door. Then he went to the desk, dropped the key there and said, "Otto, I'll give her until morning to tell us where she's been meeting him. I was banking on you bringing them in this afternoon without any of you knowing what I had in mind, because I don't want Adams to know that we know he's been using your daughter. If she tells us where he usually meets her, we'll be ready for him in the morning. You understand?"

Otto felt for the chair and sank back into it, but Beulah said, "We understand, Tom. Paw will, too, but this hit him hard. He had no idea she was meeting—that man."

"Did you know, miss?" asked Hugh Tyre.

Beulah shook her bead, then gave her shoulders a little shrug. "Well, not him, particularly, but I thought it might be a man, some man; probably one of the range riders Paw's never allowed to the house to see us."

Otto lifted his massive shoulders. He seemed to be struggling up out of a deep pit. McGrath felt sorry for him. Even Jack Kirk, who had no particularly good feelings toward Meuller, like all the other range riders in the Puma Station country of Montana, precisely because he kept them away from his handsome, big, sturdy daughters, looked like he'd like to help the older man if only he knew how.

Hugh Tyre and Frank Moore did not share this emotion. They only knew that Meuller was next to impossible to drop with a punch, and also that his elder daughter might be their key to Jay Adams. Tyre said, "Sheriff, these folks ought to stay in town tonight. As a matter of fact, Sheriff—"

"I know what you're thinking," stated McGrath. "I only have three cells, Mister Tyre, and two of 'em are full." Tom waited until he caught Otto's eye, then said, "I want you to promise me you won't go back to the ranch tonight, Otto. I want you to hire a couple of rooms down at the boarding-house until tomorrow."

"And steer clear of strangers," put in Frank Moore. "Maybe you folks won't believe it, but Hugh an' I know for a blessed fact that Jay Adams, when he's suspicious of things, will put half a dozen spies around who just sit an' listen." Moore nodded his head affirmatively. "Believe me, Mister Meuller, you're not foolin' around with some raggedy-pants horse thief."

Kirk offered to see the Meullers down to the rooming-house, which McGrath thought was a good idea. He was hungry and dirty and tired; to top it all off, his hand hurt where he'd belted Otto.

After the Meullers had departed with Jack Kirk accompanying them, McGrath pointed to the saddlebags. "You watch 'em while I eat," he said. "By the time I get back it should be dark enough to take 'em down to the church."

Outside, the day was fast fading. It seemed odd to McGrath that the roadway was serene, the pedestrians he saw abroad were simply strolling, and the town was peacefully quiet.

He went up to the restaurant, ordered a big supper

served at the café man's plank counter, and was joined there when he was half through by Jack Kirk. The cowboy straddled the counter, heaved a mighty sigh and pushed back his hat. When the café man poked his head around the curtain leading into his noisy kitchen, Jack said he'd have the same thing the sheriff was having. The moment the café man ducked back among his fiercely frying and popping utensils, Kirk said, "Say; suppose Adams has spies right here this evenin', and they saw me escortin' Otto and Beulah to the rooming-house?"

It was a possibility, Tom admitted, but not a very likely one, since Adams couldn't know yet that Helen had been arrested and had admitted that she'd been slipping away to meet Adams. As Tom reached for his coffee cup, he said, "We'll have to take some chances, Jack. If we're lucky, we'll only take the ones like this, with most of the odds stacked on our side. I don't think Adams's spies are anywhere nearly as interested in Puma Station as they are in Meuller's ranch."

After supper both Tom and Jack Kirk balanced trays of food and headed for the jailhouse. There, with Moore and Tyre looking on, McGrath fed his prisoners. The renegades ate noisily and afterwards complained about the quality of the grub. No one heeded either of them. Ruth ate a little, but Helen Meuller stood over near the back wall of the cell she shared with Ruth and stared directly at Tom McGrath, making no move toward the food Ruth set aside for her.

"Tom," she said, "he'll kill you for this. He told me if any local lawmen tried to take him, he'd kill them."

McGrath leaned on the cell door, eying Helen Meuller. "You're too nice a girl," he said quietly. "Helen, for the Lord's sake—*think*. You know what he is. You've seen enough newspapers, heard enough of the stories."

"Lies," she retorted. "He told me how every unsolved crime is blamed on Jay Adams, even though Jay Adams wasn't even in the country."

Tom stopped trying to reason with her. He dropped his glance and saw Ruth looking a quiet warning to him. He shrugged, dismissing Helen Meuller, and said, "I'm sorry, Ruth. It looks like you may have to stay another few days; maybe longer. You can't walk out of here while he's still armed and on the move."

Jack Kirk said, "Helen, you know what he'll do when he figures you're not comin' to meet him any more?"

"He'll come looking for me," she flared at the Two—Bar cowboy. "That's what he'll do. And he'll take this place down log by log."

"Naw," protested Jack distastefully. "Cut it out, Helen. He doesn't give a copper cent about you. He only figured you might know where his money was hidden on your paw's ranch, he's not comin' for you. But he'll come for your paw and your sister."

"Why would he do that?" asked Helen defiantly.

"It's right simple," replied Jack. "Because, when you don't show up, he's goin' to figure you've told the law all you know, then went into hiding until he's captured."

For five seconds there wasn't a sound. Ruth, Tom McGrath, and Helen Meuller stared at Kirk. He'd made a solid point. So solid, in fact, that Marshal Hugh Tyre stolled over into the cell-room doorway and nodded at Helen. "Miss, that's exactly what he'll do, because you see, my partner and I've been on his trail better'n a year, and we know how he thinks. He has to have his revenge for a double cross."

"But I haven't double—"

"Lady, he'll think you have. That's how his mind works. I know what I'm talkin' about. Maybe you don't care what happens to Miss Hall here, and Sheriff McGrath—but if you care at all for your paw and your sister . . ." Tyre turned without finishing and strolled back out into Tom McGrath's office.

Again there was a troubled silence. Jack Kirk walked away, too. Tom stood solemnly gazing at the handsome girl; then he too departed, locked the cell-room door, and went wearily to his desk to sit down.

"What a lousy thing we've done," he said. Then he held up a hand when Tyre would have protested. "I know, I know; we had to do it. Anyway, I think it's true. I think Adams would try to cut a throat or two if he thought Helen'd double-crossed him."

"Sheriff," said Frank Moore gravely, "you can bet money on it."

They looked at one another. It was dark out now. Their day had been eventful. As far as Tyre and Moore were concerned, it had also been fruitful. They told McGrath flatly that this was the first time they'd ever come close enough to Jay Adams to believe they might—just might—get their hands on him. They also said that the money in those saddlebags was the sweetest kind of bait for a man like Adams.

"He won't disappear this time," stated Moore. "He won't even try to lose himself until he gets his hand on this much cash."

Hugh Tyre made a smoke, went to the door and looked out, then closed the door and looked at the others. "By the back alley?" he quietly asked Tom McGrath. The sheriff nodded, got up and went over to scoop up the saddlebags. Jack Kirk grinned at the look on Tom's face. "You'll never again hold that much cash in your life," he said.

They removed the tie-downs from their belt-guns, turned down the lamp in McGrath's front office, and went out through an infrequently used rear door into the back alley where the sounds—and scents—of Puma Station came to them through a blue-black mist of night-time.

CHAPTER 12

SECRETING THE MONEY WAS RIDICULOUSLY SIMPLE, although they went about it as though expecting trouble. The little church beyond town had originally been constructed among the tarpaper shacks and dugouts to minister to the needs of the fallen men and women who had at one time inhabited those places. But that had been several decades ago before commerce and the cattle industry had brought a measure of substantial prosperity to Puma Station.

A few men still lived down there in those rotting, tumble-down shanties, but very few, and mostly they were saddle horsemen on the move, heading for another job, perhaps beginning the drift southward out of the high country. But regardless, they no longer needed the services of a preacher, or if they did need such services, they never went near the church, which amounted to the same thing.

Services were held every Sunday nonetheless, and a great many townsmen—even an occasional cowman—went down there, particularly the ones whose wives dominated their homes. Otherwise, though, the little church was left pretty much alone during the week.

The minister lived in a cottage beside his church, but

they got inside the building without awakening him, and got the money cached almost without a sound; went back outside and spread out, heading back uptown from different directions to meet at Jim Astor's saloon after making a sweep of the town to ascertain whether or not anyone had been spying on them.

Tom McGrath was satisfied they'd hidden the money unobserved, yet nevertheless he meant to have watchers patrol the town, which was the main reason he'd suggested that all of them meet at the Angel's Roost Saloon, the best place in Puma Station to recruit possemen of one kind or another.

Of course the patrons of Astor's bar knew Jack Kirk by sight. They also knew Tom McGrath, and by this time enough gossip had passed around so that they either knew, or thought they knew, who Moore and Tyre were.

The place had a fair-sized crowd of drinkers and poker players; not as many as there would have been had it been Saturday night, but enough nonetheless for what McGrath had in mind.

He had Jack Kirk recruit some men to patrol the town, and this of course aroused interest. A number of men asked McGrath if he was expecting trouble in Puma Station. His answer was brief and elemental. He told them the strangers over at the bar were deputy U.S. marshals; that they'd come here to seek out Jay Adams. And when this seemed to surprise most of the range men and ranchers, he said there was a good chance Adams or his men might try blowing the stage-company safe.

He didn't mention the church. Neither did he mention the cache Ruth Hall had brought to Puma Station with her. He had no reason, either, to mention the Meullers or the fact that he'd locked up Helen Meuller. All most of those things would have done would have been to send folks off on speculative tangents which might impair the guarding of the settlement.

Jack was pleased with his deputy-sheriff role. But he was circumspect, always, which was what had first drawn Tom McGrath to him. As Tom selected the townsmen and cowmen he wanted in his mounted posse, he'd send Jack for them. In this fashion they collaborated in acquiring six top hands. Tom knew every one of them. In fact, three of them had been with Tom and Jack in the original posse when they'd captured their "horse thief" over east of town riding a Meuller horse.

Moore and Tyre stood stonily at the bar, looking unapproachable, which in fact they were. From time to time one or the other would twist to watch McGrath at his recruiting, then face forward again. Once, Tyre said to Moore they'd made a mistake; that McGrath was a lot tougher and had more savvy than other cow-town sheriffs. Moore's answer was less concerned with Tom's savvy than with his other abilities. "I never figured his brain-power one way or another. But I sure figured his brawn-power, and if he doesn't forget our little postponed disagreement when you'n I first rode in here, I'm going to be in a bad spot. I never saw a feller who could hit that hard, or that fast." An idea struck Moore. He turned and whispered, "You reckon he'd be slow with a gun?"

Hugh Tyre, the professional pistoleer, savored his shot of rye whisky before answering, then said simply, "You're going to wind up dead if you buck him with guns, Frank. Better stick to fists." Tyre studied his companion's morose profile and shrugged. "Take your licking. You sure asked for it."

Moore turned angrily, but before he spoke Jack Kirk

appeared. “We’re headin’ back for the jailhouse,” he said. “You fellers can come or stay, as you like.”

Outside, there wasn’t much of a moon, although otherwise the ancient heavens were glowing with starshine against a purple background. The night was warm, for a change, which presaged hotter days to come. Springtime in Montana Territory was fleeting and chilly anyway.

McGrath led the way back to his jailhouse. When the four of them arrived there, Hugh Tyre asked how many men he’d recruited for his mounted posse. Tom said, “Six,” and tossed the cell-door keys to Jack. “Bring in Helen and Ruth,” he said, dropping down at his desk.

Tyre and Moore went across the room, where there was a wall-bench. They accepted the likelihood of being spectators rather than participants in whatever McGrath had in mind.

Ruth came first, then Helen Meuller. Tom pointed to chairs, and the girls sat. They were very different, but in their facial expressions there was at least a little similarity; both were apprehensive, Helen more than Ruth.

Tom dumped his hat on the desk and said, “Helen, I’m not going to beat around the bush with you. Where have you been meeting him?” Then, before the girl could answer, Tom said, “Listen to me; I understand your desire to protect Adams, but there’s a lot more at stake than just Adams. There’s your sister, your paw, every man in this room, Ruth Hall there beside you, and all the unknowns he’ll ruin or kill in the future if he’s not stopped now. I want you to figure on that for a second or two before you refuse to tell me where you’ve been meeting him. I also want you to figure on somethin’ else too; whether you tell me or not, we’ll find him. This way, we’ll do our darnedest to catch him off-guard and take him alive. The other way—without your help—it likely won't end up that way at all, because we'll just have to read sign and comb the back-country until we stumble on to him. And there'll be a fight sure as the devil."

Ruth turned to watch the other girl. All the men were also watching her. That she was troubled was obvious from the way she clenched both hands together tightly in her lap, and also by the rings under her eyes and the droop of her full, heavy lips.

"If I told you," she murmured softly, "he'd get killed anyway, Tom. He told me he wouldn't be taken alive."

"Helen; if we can ambush him; slip up and get him surrounded, and he knows there's no way to fight out, I doubt that he'll just throw himself at our guns. Men, not even men like Jay Adams, are not that foolish. Anyway; I promise you he'll be given every chance to surrender."

"Tom, I can't. I just can't do what you want. I realize we've know each other a long time; that you've always been very fair. But we're talking about the life of the man I love."

Hugh Tyre broke in quietly, saying, "Miss, we'll find him anyway. With your help or without. All Sheriff McGrath is doing, is trying to keep him from getting shot to death for your sake. Believe me, miss, when I promise you we'll get him."

The distraught girl gazed at Hugh Tyre. If she'd ever had doubts, Tyre was the kind of man to crystallize them into fear. He looked exactly like what he was: a deadly gunfighter, cool, fast, and seasoned. Frank Moore also looked the part, and he was. In the ensuing silence the weight of those two presences lay heavily. Neither Tyre nor Moore had any reason to wish Jay Adams alive. To them he was simply their current objective; they were impersonal and ruthless, which was the most deadly combination of attributes in professional killers whether within the law or outside it.

McGrath made a cigarette while he waited. Ruth Hall reached over and laid a hand lightly upon the clenched fingers of Helen Meuller, then took her hand away. It was a small gesture of sympathy and understanding.

Jack Kirk, by the door, examined the crown of his hat with meticulous care. McGrath lit up, put a steady gaze upon Helen Meuller and said no more, although his expression showed that he might.

Finally Helen faced McGrath with her head up and her chin—shaped in the same stubborn mold as her father's jaw—set hard against cooperation. That was when Tom played his hole-card, speaking before the big girl could defy him.

"If you won't cooperate, Helen, we can sweat it out of those two gunmen in the other cell. Men know a lot of ways to get other men to talk that they wouldn't use against women. Those two will talk, believe me. But then it'll be a little different."

"They don't know where he is," she said quickly. "He moved a lot before yon caught them."

McGrath nodded. "That's my point, Helen. We won't be able to stalk him and give him a chance to stay alive. We'll have to backtrack to where those gunmen last saw him, and pick up the trail from there. But they'll give us enough to go on."

Helen's mouth twitched. "Tom," she said, almost whispering his name. "Tom, promise me you won't kill him."

All the hard-faced men tensed, sensing capitulation on the big girl's part, but they showed nothing on their impassive, darkened faces. McGrath was gentle when he replied, "I've already told you that, Helen. We'll take him alive if there's a chance in the world to do it."

"But . . . if he fights you . . . ?"

"I can't answer for him, Helen; only for us. If he fights us, we'll still do our darnedest to take him alive. But none of us are going to let him shoot us."

There were hot tears in Helen Meuller's eyes; she had to fight silently for self-control, which was another indication she was coming to a difficult decision. If she found any solace at all, it had to be in the fact that she'd extracted the best terms, for otherwise those two federal lawmen sitting there like hired executioners most certainly would have shot first and lamented afterwards. That was obvious to everyone in the room. And she couldn't doubt but that these dedicated men would, as they'd promised, run Jay Adams down no matter how long it took or where the trail led them.

"I meet him three miles northwest of the ranch," she muttered, "up near the old stone trough near the hills. You know the place, Tom."

"Yes; I know it. Does he come alone?"

She nodded, looking at the hands in her lap. "Tom, if he ever finds out I told you . . ."

"He'll never find it out from me, Helen. Or from Jack."

Hugh Tyre, showing that odd sentimental quirk to his otherwise granite character, said, "Miss, Frank an' I'll tell him nothing, ever."

McGrath ducked his head to stamp on his cigarette. Adams was no fool. He'd know the minute he found himself surrounded exactly where he usually met Helen Meuller, who had told the lawmen where to get him. If Helen were thinking straight right now, instead of being all confused and emotional, she'd realize that. McGrath disliked himself and the others, but he didn't dislike what he had to do because he believed men such as Jay Adams were unfit to live among orderly folks. He raised his head and saw Ruth eying him with an expression of understanding and sadness. He ignored her to say, "Helen, you may not think you've done the right thing tonight. But afterwards you will, believe me."

That sounded so much like a Sunday sermon McGrath arose, went over and coaxed a fire in the sheet-iron stove beneath his battered coffeepot. His movements were resentful and angry. When he turned, Helen was silently crying, and Ruth had risen to stand close with an arm around the other girl's shoulder. McGrath nodded to Jack, who sidled over to open the cell-room door and take them back out of the front office.

Wien they were gone, Hugh Tyre stretched both arms and yawned, then slumped where he sat and said, "Sheriff, you got it out of her. I was betting you wouldn't be able to."

"Save it," growled McGrath. "You two want some java?"

They arose and walked over nearer the stove to accept a pair of tin cups and thoughtfully eyed the stove. Moore said, "He won't hand anybody his gun. Not Jay Adams. He'll go out with a roar."

"That's up to him," stated Tyre, holding out the cup to catch some of the lukewarm, heated-over black coffee. "But we'll see to it he gets his chance, just like Sheriff McGrath promised. You understand, Frank?"

Moore muttered something under his breath and also held out his cup.

Some time later, after the deputy marshals had departed, Jack Kirk and Tom McGrath sat drinking coffee. It was close to ten o'clock, and the town was turning quiet. Jack suggested they take a little stroll south, making certain no one was down by the little church. McGrath agreed, not because he thought there'd be anyone near their cache, but because the atmosphere inside the jailhouse was oppressive to him.

Outside, the night air restored his spirits. As they ambled along, he told Jack he thought Moore was right. "Jay Adams isn't going to surrender unless we can jump him from ten feet away and surround him with guns in such a way that he'd be committing suicide by making a battle of it."

"In his boots," opined Jack, "I wouldn't either. Why, they'll only hang him somewhere anyway, and that's no way for a man to die, like a lousy beef bein' pulled up by the throat to be gutted-out and dressed down."

"She'll get over it," McGrath mused. "The other way he'd use her and toss her away, and she'd likely never get over it."

"Right, Tom. You're dead right."

"Yeah?" growled the sheriff. "Well, I don't like playing God, Jack, now or any other time."

There was nothing at the lower end of town but darkness and silence, so they turned and ambled back, heading for their rooms at the boarding-house and some sleep. They both anticipated a hard day to come.

CHAPTER 13

THERE WERE SIX RANGE MEN IN TOM MCGRATH'S posse, plus the pair of deputy U.S. marshals, and Jack Kirk. Counting McGrath, they numbered ten heavily armed, mounted men. But they didn't leave town together. They drifted out two at a time in different directions, before the sun was up and also before most of Puma Station's residents were abroad.

It was the only way to travel in view of what they knew about the wily nature of their enemy. One of the range riders said he didn't believe they had too many men as it was, since Adams still had eight—plus himself—and since the two gunmen in the jailhouse cell had made it clear Jay Adams would, and could, hire more renegades to ride with him.

McGrath's reply to that was simple. "He doesn't believe he'll need more men—yet. And it's our purpose to make him go right on thinking that way. That's why we're going to slip away, looking like any other early-morning range men heading out to put in a day's work."

Their rendezvous point was in the forested foothills west of the Meuller ranch. McGrath and Jack Kirk, who knew every gulch and peak back there, had decided that if Adams met Helen Meuller within three miles of her father's house, his camp couldn't be very far from that trysting place. They even discussed likely places for that camp to be. But like the Southwest where water was rare, in Montana there were ice-water creeks in nearly every canyon, so an outlaw camp wouldn't be easy to pinpoint.

Still, they'd talked over the various spots and had decided the outlaws had to be in one of five different areas. Their deductions were based not only upon water at the camp-site, but also upon the availability of ample feed for that many horses, plus a proximity to the Meuller range, and finally, somewhere they could post watchers to see that no one came upon them unexpectedly.

With all those qualifications taken into consideration, the field was narrowed considerably. Another thing was in their favor which hadn't at first glance appeared even to be important: Those six range riders in the posse also knew every yard of the back-country; in late, hot summertime, cattle seeking respite from flies more than heat headed back into the shade where they'd have low limbs and brush to battle the annoying insects with, and range men had to ride all those hills and passes to find them.

Moore and Tyre were split up and paired off with local riders. Otherwise, McGrath kept Jack Kirk with him and sent the others by whatever circuitous routes they chose to follow.

The sun hadn't risen yet when Kirk and McGrath came around the shoulder of a heavy hill where they could see the Meuller place. They watched the slow activity down where men were saddling up outside the barn, and Kirk smiled. "'When the cat's away the mice will play," he mused. "If Otto was there they'd be moving a lot faster than that."

They waited, giving the cowboys time enough to walk their horses out of the yard in an easterly direction, heading slightly northeast, in fact, before leaving their shielding hill and continuing their ride toward the gloom-shadowed foothills to the west. They were the only possemen this far north. That had been McGrath's only stipulation: No one was to ride due west from north of the Meuller place on the chance that the outlaws would be able to see them coming. Even two horsemen looking like range riders could spook their enemies.

Beyond the ranch, where the land began to heave and buckle somewhat, they found trees and underbrush. Not very much of either, but enough to utilize as they progressed, should some spy be up ahead in the forest, watching, a possibility Tom thought unlikely if the camp was northward, and since Jay Adams' conceït fostered the idea that now he no longer needed to keep watch down there, since Helen would bring him all the news.

They were a mile and better from where the first row of trees marked the boundary of range and forest when Jack pointed toward a huge boulder, as large as a building, sitting by itself just before the forest began. "Ideal place for a watcher," he said. "I clumb to the top of that thing one time. There's a hole up there burnt right down into the cussed rock. Indians must've used that place for signal-fires since the beginnin' of time."

Kirk squinted at the rock. "Watchers wouldn't have to be redskins though, would they?"

"You see something?" asked McGrath, concentrating on the huge stone.

"Thought I did," Jack answered, and shrugged, turning his gaze elsewhere. "Probably just a shadow. That's the trouble with ridin' out before sunup."

McGrath kept studying the immense boulder. They dropped into a gulch and emerged upon the far side, passed through a little stand of blackjack-oak and startled a slumbering buck from his bed. The deer exploded into a bounding race to get away. He didn't look back even once to see whether the mounted men were after him or not. They weren't.

Closer to the stone, McGrath softly said, "Jack, split off. Ride around the rock from the right. I'll take the left."

Instantly Kirk was alerted. "You see someone?"

"Sure did. He was atop the rock; now I think he's behind it. Probably got his horse around there. Split off. If he shoots, he'll get only one of us."

"That's a happy notion," muttered Kirk, reining

away.

They didn't see the shadow again, but obviously, if he was indeed back there, he'd know he'd been spotted by the way those oncoming riders had split up, approaching him from two divergent directions.

He was back there all right. So was his horse. But when McGrath was close enough to use a carbine, the watcher reined out atop his animal and awaited their coming. It was one of the possemen.

Jack got to him first and was saying he'd given the later arrivals a scare when Sheriff McGrath also came up to them. Tom asked the posseman where the man was he'd ridden out of town with. The posseman turned and pointed back in the low-down fringe of trees. "Watering his horse and waitin'," he answered. "Come on; I'll show you."

That big rock was not the rendezvous place, and Tom McGrath didn't like the idea of four mounted men riding together through the forest; they'd make too much noise and they'd be too easily discernible even though, with no sunshine yet, the world of trees and underbrush through which they now had to travel was still speckled and gloomy.

When the three of them reached the place where the fourth posseman was resting his animal, this cowboy said he and his companion had scouted through the southerly foothills and hadn't even encountered an old camp-site. McGrath was only slightly interested. He'd never thought Adams would be southward.

He led his companions straight west through the forest, across a little grassy clearing where bluejays came down to the lowest limbs and shrilly scolded them for intruding, and down into a broad, protected valley where, through trees, they caught sight of an old rotting log house with loopholes in its walls and a mud-wattle chimney standing as erect as the day it had been built.

There were several horses visible down there, browsing among the trees, which weren't very thick in that place, probably because some long-dead trapper, who'd made that house, had burned them to keep alive in Montana's severe winters. This was the rendezvous. It was hidden adequately unless, of course, Adams was keeping an intense watch down here. With the exception of perhaps Moore and Tyre, all the other men had known where the cabin was; had, in fact, camped either inside it—if the weather was cold or rainy—or beside it next to the creek while searching for cattle each autumn.

As they rode in, Moore, Tyre and two other men came to the doorway to eye them. All but two of the possemen were now here. McGrath asked them about those other two, but none of the men at the log house had seen them.

They waited, with the chill of a forest making them go inside. They smoked and idly discussed what lay ahead. McGrath's plan was for them to continue north as they'd gotten this far, in pairs and riding different trails.

"But from here on," he emphasized, "you're likely to get shot out of the saddle any time, so you'd better use both eyes all the time. And don't do anything if you find them. No shooting unless they start it. Just come back and hunt up the rest of us."

Jack said, "How about that place where he met her?"

Tom nodded. "I'm coming to that. All of you except the federal men know where that stone watering trough is. Make your sweep toward that place. I know the trough sets out a short distance from the forest, but you fellers stay in among the trees. Jack and I'll head straight for the trough. If we're lucky, we'll bring this off without a shot."

"Ain't likely," remarked a laconic range rider.

Tom agreed. "I reckon it isn't very likely. But I've got my reasons for wanting it done this way. If you see Jack an' me out there alone, just keep watch. If you see a stranger out there, try and get around him. If he's left his horse tied in the trees, try an' cut him off from it. But don't shoot."

The last two possemen rode in. Tom went over it all again for the benefit of those men; then they walked out to their horses, as silent as fugitives, mounted, and came together in front of the old cabin where one man asked a question: "Tom, where will Adams' men be? I figure the one you're hopin' to catch by the trough will be Adams himself, an' that's fine. Only I keep wonderin', if there's gunfire, where those other fellows are likely to hit us—from the north, or from behind us to the west?"

McGrath had no answer, only a guess. He said, "Adams wouldn't have his whole crew spying on him when he's with—someone—by the trough. If there's gunfire, you'd better watch in both directions. But my personal guess is that the camp's on north where they could light out if they had to and get into the wilder mountains that curve around back there. Let's go now; and be careful."

The men milled a little, seeing which way the others were heading out before taking different tangents leading in the same general direction. Jack Kirk, heading back down toward the fringe of the forest with McGrath, watched until they were hidden by tree-trunks from the old cabin, then squared up in his saddle and said, "Good men. Even those two marshals seem to know what's goin' on."

McGrath had no comment to offer for the obvious reason that Moore and Tyre had to be good men on the trail to have lived as long as they had, while wearing badges. He was thinking ahead; there was no guarantee Adams would be at the meeting place. On the other hand, the composite picture Tom had erected in his mind of this notorious outlaw made him think Adams *would* be there.

All men had vanity. Some simply had more than others. But a man like Jay Adams, who had eluded every posse ever sent against him, who had broken the law with total impunity when he'd chosen to do so, and who was not only ruthless, but who was also quite a ladies' man, had to possess more than the usual amount of masculine vanity. He'd be at the trough if for no other reason than to reaffirm to his own satisfaction that he was irresistible to Helen Meuller.

McGrath was banking on the probability that this glaring weakness was going to be the factor which finally caused Jay Adams' downfall. He hoped Adams wouldn't force them to kill him; not solely because he'd given his word to do everything possible to take the outlaw alive, but also because he wanted to see, and hear, the most famous of all bank and train robbers of his generation.

The rewards for Adams totaled a very respectable sum, but that didn't move Tom McGrath much. He wasn't a greedy or, for that matter, a very ambitious man. What he wanted was to get Jay Adams out of his country; he wanted to see the money handed back to its owners, and see his bailiwick return to normal again.

Jack said, "Tom, just how good is Adams with a gun?"

"Tyre says he's very good. Ambidextrous."

"What?"

"Just as fast and accurate with one hand as with the other one."

"Oh. Why not just say two-handed?"

They were passing through a particularly dense patch of forest not more than a mile from the stone trough now, with the sun peeping over the jagged rim of the eastern world. It was cold in there, and gloomy. Even after the sun got all the way up, its light still couldn't make much headway against all the tree shadows. Ferns grew hock-high, indicating there was sub-surface water. Their horses didn't make a sound, plodding over pine needles and fern fronds. In a way, it was eerie enough to make their nerves crawl.

CHAPTER 14

THE FIRST THING MCGRATH NOTICED WAS THE absence of birds. The second thing he noticed was the trough out there with the sun striking it, making it look less ancient and grey. There wasn't a soul beside it. Jack whispered a comment about that. McGrath dismounted back a short distance, tied his horse and stood beside the beast's head, gazing around.

"Too early yet," he said, answering Jack's obvious thoughts. "That was the idea. Get in here and get into place before he came along—if he comes along. Stand at the head of your critter, Jack. He'll hear Adams coming before we will. Keep ready to pinch off any nicker the horse might make."

They stood in the chill and gloom for a long while. Eventually the air warmed up somewhat, although sunshine still couldn't break through the eternal gloom down among the giant trees.

A few curious little wood birds came back to cock their heads from the high branches and speculate on what those earth-bound animals down there were doing, standing like statues in the forest's hush. A raffish crow, evidently having spent the night in the forest, came noisily down through the trees, keeping up his cawing, and passed the motionless rnen without even seeing them. At once two much smaller birds, who probably had a nest close by, went after the crow, flying above and diving to peck at his head. The crow rolled and flapped and scolded, but they persisted. This was the only thing McGrath and Jack Kirk had to watch, so they kept observing the running fight until the crow suddenly heeled over, heading for the nearest trees where he could lose his pesky tormentors, complaining loudly every inch of the way.

A coyote pup yapped nearby. The horses spotted him first. He wasn't more than five or six weeks old and still had the downy, colorless fuzz he'd been born with, showing in ragged tufts through the longer guard-hairs of his natural fur. His tail was full and nearly as long as his body. He'd run a few yards, sniffing for grubs, exploring old golpher warrens, yap, then trot on again. When he came straight for the men, one of the horses stamped. The pup stopped, stiff-legged with alarm, sniffing for some clue as to the whereabouts of these unwelcome trespassers. Even when he finally made out both men and both horses, though, he didn't turn in panic; like all coyotes, adult or pup, he had an insatiable curiosity. The men eyed him, and he returned their regard, wrinkling his nose at them to catch every scent. Finally he turned tail, trotted off perhaps fifty feet, turned back and sniffed some more. Then he yapped again and shuffled off, not overly concerned that the most deadly of all enemies of his kind were watching him.

Jack Kirk looked over at McGrath as though to say they'd stopped in a populous place. Tom gravely inclined his head and went right on waiting. He had the patience of an Indian. He also had the acute sensitivity of one, which any professional manhunter either had to be born with or acquire if he were to be successful at this kind of thing.

They heard subdued, abrasive sounds off to the west and heard the woods back there go suddenly quiet as though the animals and birds had spotted fresh intruders. McGrath was confident these would be possemen, but he didn't have much time to speculate, for his horse abruptly lifted its head, little ears cocked forward, and peered up-country toward the shadowy north.

McGrath slowly raised a bent hand, ready in an instant to cut off at the nostrils any sound his horse might make. But the animal remained silent and alert, concentrating on whatever sight or scent had captured his interest. McGrath glanced over. Jack stealthily cupped a hand around one ear and pointed. Tom inched ahead slightly to see up through the trees in the direction Jack had indicated.

He didn't see the rider until he heard a shod hoof scratch across rock. That pinpointed the location of the rider for him. When the man was visible Tom studied him for a long while. The rider was a lean, rough-looking man dressed in a slovenly, haphazard fashion. But he had a Winchester slung forward of the saddle-swells with the steel butt-plate hoisted up to within easy reach of the man's free right hand. He also had a six-gun tied low on his right thigh; an old gun, but it glistened in the greeny, uncertain light with fresh gun oil.

McGrath had never seen Jay Adams in his life except in newspaper photographs, but he knew even before he saw this man's face he couldn't be the notorious outlaw.

That he was a wanted man wasn't disputable; no men but those who lived by guns carried his weapons as the stranger did. But he was too unwashed, too hatchet-faced and slovenly to be Jay Adams.

Tom turned. Jack Kirk was intently watching the stranger's approach, his right hand lying lightly against his hip holster. Whatever Jack might have thought of that renegade personally wasn't influencing Jack's determination, if the moment came when he had to use his gun, to use it fast and accurately.

The man stopped and turned in his saddle, looking all around, but didn't act the slightest bit apprehensive; he rather did this as though it were a habit of long standing. He was down near the foremost fringe of trees by that time, a considerable distance from McGrath and Kirk, and with sunshine filtering through to mottle him.

The horse he rode was well built, with powerful forelegs and muscled hams. He wasn't a colt, but neither was he a smooth-mouthed animal. McGrath judged him to be about seven years of age, in his prime, and as tough and durable and fast as an outlaw would want a horse to be. Also, the horse was quiet; when the man stopped him to scan the forest, then also study the open country beyond the trough, the horse stood patiently, awaiting his next command.

The outlaw cleared his throat, spat, then dismounted. Tom looked back. Jack Kirk was leaning into a slight crouch. He wasn't paying the least attention to anything except that renegade down there.

The little coyote yapped somewhere off to the west, then barked at something, probably a stolid old porcupine or an equally slow-moving, indifferent varmint who neither could, nor would, try to get away from him. That was the only sound. The outlaw turned back and scanned the forest briefly, then went ahead, tied his horse to the last tree west of the trough, and with agonizing slowness built himself a cigarette while the lawmen watching him grew damp in the palms. He lit up eventually, exhaled a cloud of grey smoke, and stepped over to the very edge of the forest, stood a while making another careful study of all the open country beyond, and finally walked out to the trough. There, sunlight hit him, showing his narrow, cruel face to be dissipated, loose and alert. The man had a wild-animal wariness even when he stood, as he now did, feeling no peril at all. He smoked, continued to gaze around, and finally concentrated on just peering south, which would be the direction from which Helen Meuller would approach.

Without any communication between them, it was clear to both Kirk and McGrath what the man was doing. For some reason he'd been sent to intercept the Meuller girl at the rendezvous. Whether that meant Jay Adams was on the move again, or what it meant, neither McGrath nor Kirk could know, but they were aware of the significance, which was simply that Adams in all probability suspected nothing; otherwise he wouln't have risked losing one of his renegades. There was a very thin possibility, too, that Adams suspected something and had sent the man out there to see what would happen.

It was this latter thought that held Sheriff McGrath

thoughtfully motionless for so long. He could try taking the outlaw, with or without a fight and gunfire, or he could let him go. If Adams was using the man as bait, he'd be wise to stay concealed and let the outlaw eventually abandon his vigil and go away. If the man was simply a bearer of some message, then he probably should be taken.

The key to the riddle for Tom McGrath wasn't so much whether they could capture the man without gunfire as it was what they might learn from him; whether they could force him to lead them to the outlaw camp.

Jack turned his head slightly and gazed at McGrath. Without seeing this, McGrath knew, because he'd been standing motionless and undecided too long. He made his decision, lifted out his .45 and softly stepped around the head of his horse to begin stalking through the forest gloom toward the trough. Once he paused, with a huge old rough-barked red fir in front of him, to gaze back and see what progress Jack was making. The Two-Bar cowboy was being even more careful not to make a sound. He was fifty or sixty feet farther back.

McGrath resumed his stalking. Out in the sunlight, the outlaw smoked and relaxed; he studied the land in all directions to pass the time, but he appeared most interested in the southward range.

McGrath got down near the final fringe of forest. Here, sunlight reflecting inward from out in the open country diluted all shadows. McGrath had the outlaw's lanky form ahead and slightly to his left when he raised his six-gun, placed a thumb pad firmly over the hammer ready to cock the gun, and said, "Mister, if you so much as move an arm, you're dead."

He didn't say it loud enough for anyone but the

outlaw and Jack Kirk to hear it, but it was loud enough in that renegade's brain to stiffen him temporarily, his right hand with the cigarette in it suspended in mid-air between waist and lips.

McGrath said, "Jack, go get his gun. Don't get between us."

Kirk moved swiftly through the last few trees, approaching the outlaw from the far right. When he was close he muttered something. The outlaw turned his head the slightest bit, looked at Jack, then dropped the cigarette into the trough and put both his hands chest high. Not until then did Kirk move in.

But the moment Jack reached for the outlaw's gun, the man whirled with blurry speed, caught Jack by the shirt and tried to swing him between Sheriff McGrath and himself. Jack was strong and wiry. He dug in both high heels, swung at the outlaw's face with his left, and clung stubbornly to the man's holstered weapon with his right hand.

The outlaw turned his head sideways to roll with that left-hand punch, then clawed savagely at Kirk's restraining hand upon his holster. McGrath sprang ahead, covering the ground in lunges. He caught the outlaw desperately fighting to throw off Jack in his direction, side-stepped and pushed his pistol into the man's face. Then he cocked it.

All the fight ran out of the renegade. He was as near death as a man could be, and he knew it. Jack knocked off the man's hand from his shirt front, called the outlaw a fierce name and wrenched away his six-gun. Then he stepped back as though to throw his Sunday punch, but McGrath grabbed their prisoner, turned and hurled him violently toward the forest.

They took the man into the shadows again, where it

was cooler but where, also, no one would inadvertently spy them, and ordered him to sit down. The man put his whole attention upon McGrath, who wore a badge, and scarcely looked again at the Two-Bar cowboy.

"What are you doing here?" asked McGrath, easing down his hammer before holstering the six-gun.

"Doin' here?" growled the captive. "Why, I was just passin' through, seen that old trough, and walked out there to get a drink."

"Then why didn't you drink?" exclaimed Jack Kirk bitterly, stuffing in his shirt and looking disgruntled.

McGrath leaned down, caught the man's shirt collar in his fist and cruelly twisted it tighter and tighter. As the outlaw's breath was cut off, he reached for McGrath's forearm with both hands, trying to tear free. He didn't succeed in even loosening the stranglehold.

McGrath suddenly threw the man back down, releasing him, and said, "Once more, mister; then I'm going to overhaul you: What were you doing out there?"

"Waiting," gasped the renegade, sucking in big lungsful of air.

"Waiting for Helen Meuller?" asked McGrath.

The outlaw's gold-flecked, close-set eyes shot McGrath a quick look. He started to swear quietly. He said, "I told him and told him you couldn't trust one of 'em. I tol' him that every lousy chance—"

"You answer the question," snapped McGrath. "Were you waiting for Helen Meuller?"

"Yes. And that's all I'll tell you."

McGrath reached for the shirt-collar again, but this time the outlaw twisted swiftly away—and walked straight into a fist thrown by Jack Kirk. The man flopped back upon the pine needles rapidly blinking from that solid strike.

Jack said, "Sit up, mister. That wasn't a hard one; just a promise of what's next. Sit up an' answer the sheriff's questions!"

The outlaw sat up, but he was groggy and remained like that for some little time. Without being asked again, he said, "All right; I was waitin' for the girl to tell her she wouldn't be able to meet her boy friend today; that he couldn't come."

"Her boy friend," spat out McGrath. "You mean Jay Adams, and he's not her boy friend. Why couldn't he meet her?"

"He ain't around, that's why."

"Where is he?"

"I dunno."

Jack cocked his fist; the outlaw saw it poised and ducked his head. But Jack didn't fire the blow because McGrath raised a hand. Tom had heard something. He pointed west up through the trees. "Go see who that is," he ordered in a low whisper, and drew his gun, dropped to one knee, cocked the weapon and pointed it at their prisoner. "Mister, you so much as breathe loud, and I'll blow your head off."

Jack sidled over behind some trees on their left and began slipping ahead deeper into the forest. Whatever sound Tom McGrath had heard had, in all probability, been made by the converging possemen, but neither McGrath nor Kirk wanted to take the chance it might not be.

The outlaw craned a look over one shoulder, then straightened back, facing McGrath. He began to smile wickedly. Obviously, he thought whoever was sneaking down on them was some friend of his. But he obeyed Sheriff McGrath; he didn't say a thing or make a sound.

CHAPTER 15

IT WAS THE POSSEMEN COMING ON FOOT AND WITH carbines. Jack brought them. As the prisoner looked at all those men with Winchesters converging on him, his eyes flicked over their faces, their weapons, and finally considered their numbers. They drew up around him, looking down dispassionately. He swallowed uncomfortably.

Jack said, "Wrong guess, mister. It wasn't your friends after all, was it?"

One of the possemen said, "Is that Jay Adams? I had him pictured as a heap different from this skinny weasel. Why, this feller doesn't even wash."

McGrath was putting up his six-gun. As he turned back he said to the prisoner, "What's your name?"

The man's answer was sullen and spiteful. "Tom Grant."

Hugh Tyre, looking down his nose icily, said, "Why not Abe Lincoln? That's a well-known name, too. Or Robert E. Lee, or Rutherford B. Hayes?" Tyre turned toward Moore. "Frank, he doesn't want to cooperate." As though by order, Frank Moore leaned his carbine against a tree and moved closer. Tom McGrath made no move to prevent what was obviously going to happen. The outlaw saw Moore coming, saw the look in his eyes and the thickness of his body. McGrath, watching impassively, let Frank get close enough to reach out, then said, "Where is Adams?"

The outlaw's head jerked back around. He was sweating, although it wasn't even warm in the forest. "I told you, I dunno. He said for me to come down here'n meet the girl; tell her he'd make it tomorrow afternoon but he couldn't get here today."

"Did he ride off alone?" asked McGrath.

"I can't even tell you that, because he was still in camp when I left to come down here. And that's the plain truth."

Hugh Tyre knelt beside the outlaw. "Where is the camp?" he asked.

The outlaw looked up. Frank Moore was poised to drop on him. Hugh Tyre's quieter, less obvious intentions were nonetheless unfriendly. That much showed in the intent, merciless stare of the older federal lawman. The outlaw squirmed, looking back at Tom McGrath. There was no mercy in that face, either. "Back in the woods," he mumbled. "North o' here in a big clearing."

They could hardly hear his words, and afterwards the outlaw stared at the ground. Tyre rose. Moore backed off, straightening up with an expression of satisfaction across his face. The other possemen looked pretty much the same way. Tom sent Jack Kirk after the outlaw's mount, and they all stood there in an armed circle.

The outlaw said, "Listen; I'm tradin' you that information to get loose."

Frank Moore said a harsh word full of scorn. None of the others even bothered to say that much. Jack brought back the horse. "You aren't goin' to put him astraddle it, are you?" he asked McGrath. "Tom, in all this darkness among the trees, he could make a break for it and be lost in a minute."

Tom turned to the others. "Fetch your mounts and come on back here," he said. "No noise." As they departed, he looked across at Jack. "We're going to need this horse; that's why we'll take it along. But Mister Grant isn't going to ride it; he's going to walk."

Grant stood up, dusted himself off and felt his face where Kirk had struck him. He watched the others disappear in among the trees, looking murderously after them; then he slumped there and considered Kirk and McGrath. "If you're figurin' on a raid," he told them waspishly, "you'll never bring it off."

McGrath was interested, although Jack Kirk looked skeptical of whatever the captive said. "Why not?" McGrath inquired.

"Because we got watchers out all aroun'."

McGrath had expected nothing less and said so. After that the renegade seemed less certain of himself and more apprehensive for his friends. Then he tried a fresh tack on McGrath. "Listen, Sheriff; I can take you in there without any o' us bein' seen. Otherwise you'll never even get close."

Tom nodded dryly. "Another trade?" he murmured, and when the outlaw bobbed his head up and down McGrath shook his head. "No trades, Mister Grant, and no tricks. We know this country better than you an' your friends do. As soon as we're heading in the right direction, we'll figure out which big meadow it is. After that we'll also know how to get in there. But if you get cute along the way you're goin' to get a split skull or a bullet. Remember that."

Jack Kirk had been figuring. Now he came up with something. "Guessin' those renegades of Jay's got up like most folks, around sunup, and guessin' this one ate before he came down here, Tom, would make it maybe four miles to their camp. Now I know that north an' northwest country, an' the only meadow back in there big enough to feed all their stock and still have decent places for watchers to keep guard is the place called Mormon Flat."

Tom turned as the others began coming down through the trees. He said, "Get your horse, Jack; then keep an eye on Mister Grant while I get mine." He made no comment at all about Mormon Flat. He didn't have to; the look of wariness with which Grant had listened to Jack had fairly well confirmed Jack's contention that the outlaws' camp was there.

They all got astride except the prisoner. He turned sour and sullen when Frank Moore dropped a loop around his neck and picked up the slack with a little smile playing around his coarse lips.

"If you sing out," said Frank, giving the rope a little flick before settling one dally around his horn, "or if you start kickin' stones while we're moving, I'm going to toss this rope over a low limb and haul you up sure as you're standin' there. It'll be an accident, Grant; you'll be running away an' the rope got tangled in some trees, stranglin you." Frank's smile turned downward. "If you think I'm bluffing, just make an unnecessary noise."

Jack and Tom led off, back-tracking the outlaw, which wasn't difficult in bruised pine-needles and ancient dust. His sign went along through the lower fringe of the forest as though he'd purposefully ridden down there like that in order to be able to see out across the open range beyond.

The tracks went straight north for a mile or more, then angled inward, deeper through the trees and across several little rocky gullies. Here the possemen had to be extremely careful not to dislodge loose boulders or make any more noise than they had to. In the trees again, though, those layers of rotting needles muffled nearly all sound.

The man calling himself Tom Grant kept up well for the first mile. During the second mile, where the country got rougher, rockier and more broken, they halted twice to let him catch his breath. The up-country of Montana is in high attitude. Even for men conditioned to it, that thin, fragrant air isn't sufficient.

It was at the second stop that McGrath left them, scouting ahead on foot, following the marks left by Grant. He went almost another mile ahead before he saw anything. Then it wasn't a man on a hill or perched in some vantage-tree; it was a scruffy old bone-rack bull bison snuffling around where seepage water kept perhaps an acre or two acres lushly green.

The bull was old; no doubt he'd lost most of his teeth. That could be a catastrophe for buffalo, since they only had lower teeth anyway and he was as scarred and stiff as any old battler his age would be. Also, he was either hard of hearing, which wasn't too unusual, or else he was making enough noise eating to miss the sounds of the man watching him.

Like all buffalo bulls, he was huge and vicious-eyed, with no fear of any enemies except men or wolves, and he moved with the ponderous, stolid, stupid sluggishness of his kind.

He was dangerous, even though so old he was stiff, and in among a forest fall of trees, he would charge a man. Tom felt no apprehension on this score; he could drop the old bull before there was any real personal peril, but if he fired his six-gun to do it, it would be the same as announcing his arrival in outlaw country.

He pondered the dilemma from beside a lightning-struck dead pine tree. Grant's trail led straight across that green place, obviously, because the canyon began to pinch down on both sides in here. Pelting the old bull with stones to make him move wouldn't cause anything but resentment. Shooting was out of the question. He studied the land form beyond. The canyon appeared to widen and bend around to the north as it did so.

This particular canyon wasn't familiar to McGrath, but the general flow of the land was. He was still in the low-down foothills, but within another mile he'd be hitting steeper climbs and rougher going.

He moved around the seepage spring and the rheumatic bull bison. It was slow going, underbrush tore at him, sweat ran into his eyes, and the tension made his belly knot up. But he had a plan.

When he was over the hump and down into the canyon up where it widened, he could see behind him up there that a clearing showed. A little hurrying whitewater creek crossed that clearing, making a chuckling sound. Between McGrath and the waiting possemen was the ancient bull buffalo. McGrath considered which way he'd go when the time came, picked up several rocks and heaved them one at a time.

The first stone struck mud, making the old bull blink and toss his head. The second one was even closer. The bull lifted his mammoth head, beard almost reaching the ground, and turned muddy little savage eyes all around. The third stone bounced off his sunken ribs, making a drum-like sound, and although it couldn't have pained him, nevertheless it startled him. The old bull switched his little broom of a tail, showing the rise of his unreasoning wrath.

McGrath had the range and dropped several more rocks upon the big animal. Then he stepped out from behind a tree and waved his arms. Finally the bull saw him. His little tail stopped moving, that gigantic head dropped slightly, and the buffalo backed up one step at a time until he was out of the mud and wet grass. As soon as he felt solid, stony ground he pawed, first with his left hoof, then with his right one. He was now ready for combat. But he didn't charge; instead, he walked with formidable deliberation around the mudhole, keeping those tiny bloodshot eyes upon McGrath. He didn't even hasten his walk into the familiar side-to-side shambling trot of bison until he had the seepage spring behind him and only Tom McGrath ahead.

The stupidest large animals on the North American continent, buffalo were nevertheless savage fighters, and to an unarmed man, nothing was more deadly. In their tiny brains they had learned over the millennia that man was their foremost foe. Indians since the Stone Age had survived only because of bison; the entire red-man economy and culture had been built around buffalo. With the advent of the white man and his "thunderstick"—his gun—the bison had simply turned from one two-legged predator to another.

This old bull was no exception, and probably in his long life had charged his share of men. He had scars on both sides as well as on his huge curly head which could have been the result of bullet wounds or arrow wounds. Tiny brain or not, after thousands of years, even the least intelligent of the big beasts came to equate danger with two-legged critters. This one never deviated, never shifted his glance. He was concentrating only upon smashing Tom McGrath, then pawing him to a red froth.

Tom let him begin his trot before he moved back around a tree, running always northward up the widening canyon. From time to time he'd let the old bull see him, switch its tail and start forward again. Tom also pelted the huge animal with stones until the buffalo was in a murderous mood. Then Tom, panting hard from all this exertion, barraged the bull from dead ahead, and afterwards dropped down into a thick growth of underbrush and scuttled up the slope out of harm's path.

He'd decoyed the old bull away from the narrow part of the trail, which he'd wished to do, and now ducked around after the animal went trotting past, slavering and grunting to himself, heading back down the canyon to his companions.

Tom had to stop several times to catch his breath. When the others finally saw him coming, they ganged up, full of curiosity. He explained about the buffalo and paused to rest a moment, then climbed into the saddle, flung off sweat, and led out.

They went carefully. By now, of course, the old bull had decided the two-legged thing had escaped. He might spend as long as it took for his wrath to diminish hunting for Tom, and he might also amble back down where the seepage spring was. Jack Kirk saw the size and depths of the tracks and said softly there hadn't been any buffalo in the Puma Station country since he could remember. Hugh Tyre, older than most of the other possemen, seemed to know more about bison than the others. He said if that bull, old and stiff-jointed as McGrath had said he was, came down this canyon and saw the mounted men there, he'd charge them sure, and they'd have to shoot him, because it wouldn't be possible for all of them to get out of this narrow place before he caught them.

But as Tom led them around the bend away from the seepage spring, with every eye probing ahead and several hands lying ready on gun-butts, there was no sign of the buffalo. Evidently he'd decided to amble on up the trail in search of a less popular place to eat.

Grant stopped when they got up into the widening

trail almost to that first empty clearing where the creek was, and ran a slow, scowling look after the old bison's tracks.

Suddenly, somewhere up ahead, a man let off an ear-splitting howl, and gunshots erupted. Horses neighed in panic, and an enraged bull buffalo let off a tremendous bellow of pain and rage.

Grant said, "Looks like he charged the camp!"

Tom stepped down, drew his carbine, motioned to the others to come along, and told the nearest posseman to remain with the horses and the prisoner. The others all piled off in an equal hurry, taking their Winchesters with them.

Up ahead, the shouting and shooting continued. Then it stopped as suddenly as it had begun, silence reigned briefly, and a man profanely called on other men to go make certain none of the horses had stampeded out of the country when that buffalo had charged the camp.

CHAPTER 16

THEY SKIRTED THE LITTLE CLEARING WITH THE CREEK in it, plunged into more forest with the ground turning more rocky, slowed where McGrath held up a hand to them, and finally, with no more noise coming from up ahead, halted altogether.

Tom told Moore and Tyre they'd better wait a moment; had better give the renegades up there time enough to recover from their mishap.

"They're all armed right now and ready to shoot," he said. "Give them five or ten minutes. This'll pass."

Moore wasn't anxious to halt. Neither was Hugh, but at least Tyre was old enough to have patience. The other possemen were simply wary and watchful. Jack Kirk said the Mormon Flat meadow was about a half to three quarters of a mile ahead through the forest. This inevitably brought out a question. If the outlaws up ahead were hunting loose horses, and came down that way perhaps heading for the little meadow farther back with the creek in it, what should they do?

The answer was of course elemental: They shouldn't be in sight. Tom led them over to the east a short distance where an enormous old deadfall lay; its root structure, where it had been wrenched from the ground, stood taller than a mounted man. There, in position to prevent outlaws from streaming southward where they'd left their mounts and their captive, they squatted in uncomfortable silence.

Hugh Tyre seemed unruffled. Frank Moore, on the other hand, was scarcely able to restrain himself. He kept fondling his carbine and peeking around the deadfall as though expecting to catch sight of a two-legged target.

They waited for nearly fifteen minutes, sweating and tense; then McGrath took only Jack Kirk and passed noiselessly through the trees on a short scout.

It turned out to be a longer search than McGrath had anticipated. They smelled dust in the still, heavy air, and heard horses snuffling as they gazed, but Mormon Flat wasn't just beyond the next tree; only the clearness of that mountain air had made it sound that way.

They came upon the old bull's tracks, passed a small oak which he'd broken, probably leaning into it to brush off tantalizing flies; and when Jack caught the scent of tobacco he brushed McGrath's arm and dropped down in the heart of a spiny bush, pointing.

The same creek which raced across the smaller rear

meadow also ran up through here. They couldn't see it, but both knew where it was because both had been there before. McGrath told Jack to cover him and belly-crawled with his carbine in the crook of both arms until he'd covered three hundred painful feet in this fashion. There, beyond the farthest tree, he caught the hard, brittle flash of sunlight off a clearing. He also heard men loudly, angrily talking. He went another fifty feet and rolled into a soft little shallow swale where he had a good sighting on ahead.

The old bull was down out there, evidently dead, and several men, some hatless, several still wary around that huge carcass, were gazing at the buffalo. What had occurred was simple enough to guess. When Tom had led the old bull north to get him away from the seepage spring so the possemen could ride on through, he'd ambled along after losing Tom until he'd picked up the fresh scent of other two-legged critters. With his boiling wrath still high, he'd then gone in search of these fresh enemies, had seen them through the trees in their camp, and had burst upon them with a roar of fury.

It wasn't difficult, either, to imagine the astonishment of those lolling outlaws out in the clearing when they heard that bellow and saw the scarred, huge old gaunt bull buffalo charging them out of the trees.

Tom lay another ten minutes getting the position of the camp, then turned and wiggled back to where Jack was uneasily waiting. Together, they then undertook the return trip where the others were also fidgeting and anxious.

The sun was well up; in places its light came down through the tall trees, making a diffused, cathedral kind of soft light, sometimes golden, but just as often bluish or greenish. The coolness persisted, but with overtones of high humidity which made it almost hot anyway.

There was dust everywhere; in the forest, in the air, underfoot. The place had a musty, spicy fragrance to it which none of the possemen heeded as McGrath and Kirk glided back around the big old deadfall to them and Tom related what he'd seen up in the big clearing.

Hugh Tyre, impatient at all the delay, proposed that they move at once up through the trees. Tom didn't argue against it. He'd seen the camp up there and thought the outlaws had recovered sufficiently to be off-guard. As he said this, a posseman stood up from behind the deadfall to look innocently around. They all heard the man grunt, stiffen for a second, then go for his six-gun.

From beyond the tree a shot rang out. None of the others could see the shooter out there. In fact, it happened too quickly. One minute the posseman was looking around; then he dropped like stone with a bullet in his head.

At once the others boiled out of their hiding place. Tom saw a fleet movement over through the trees to the east of them. He also saw a horse standing where someone had dropped his lead-rope, looking frightened and bewildered.

A yell was raised as the fleeing man cried out the alarm to his friends up in the big meadow. Instantly, other shouts were raised. Hugh Tyre swore with angry feeling and started running forward. The others followed Hugh's example, heading in a ragged line for the clearing up ahead some little distance.

Tom swore to himself; none of them could cover all that ground swiftly enough now to catch the outlaws unprepared. As possemen streamed around him, Tom looked back to where they'd left the dead man. It had of course been a coincidence; the posseman had happened to stand up for a look around at the precise moment one of the outlaws, evidently returning from the smaller meadow with his frightened horse, had also happened to glance over at the deadfall.

Of course what hadn't been a coincidence was the fact that the outlaw had been twice as fast and accurate as the dead man, but then no one ever expected everyday range riders to be as quick with their guns as men who lived by weapons.

Frank Moore got ahead of Hugh Tyre in the race through the trees. Tom heard Tyre yell a warning to Moore. It was in the nick of time. Up front, a volley of irregular shots erupted. Apparently the fleeing man had been able to warn his friends from which direction peril was approaching. The bullets cut around where Frank had been, but after Tyre's shout Frank had ducked behind the big red fir tree.

The other possemen also dropped flat or jumped for cover among the trees. Jack Kirk threw himself into a heavy shrub of some kind, and Tom heard him groan with profanity when the thorns dug into him.

The outlaws were shouting to one another. Most of it was indistinguishable, because they were simultaneously raking the southward forest with gunfire, but McGrath heard enough to understand they wanted to get saddled up and away from the meadow.

He'd seen some of their horses out there, along the westerly edge of the meadow, when he'd reconnoitered the camp earlier. Now, catching Hugh Tyre's eye, he jerked his head and began ducking and weaving his way in and out of trees, trying to get around the scene of hostilities to the west. Tyre either second-guessed McGrath's purpose, or joined him out of a curiosity, and also left his position to run west.

The fighting swelled steadity in among the trees. It was the clear intention of the outlaws to keep the possemen out of their meadow. It was just as grimly the intention of the possemen to try and force the outlaws to drop back out there where they'd have good sighting at them.

Tom didn't look back as he trotted around the big curve of the meadow. In size, that grassy place with the creek running across it had to be at least eighty acres, perhaps more. To run around a perimeter that large took time. Tom pushed himself hard, unwilling to stop even to catch his breath. Once, where the land dipped into a sort of brushy trough, he looked back. Marshal Tyre was coming, but he was slowing considerably, too. The distance between them gradually widened until McGrath, believing they might be far enough around, stepped up into the gloom of a giant pine and stopped.

By the time Hugh Tyre got up there, too, they were both panting. Tyre leaned his carbine against the rough bark, looked out, drew back and shook his head at McGrath, not even wasting a breath speaking.

Tom said, "Wait," and sidled around their tree, heading east toward the big meadow from this fresh direction. He could see the sunshine out there even though he was easily two hundred feet from it. Closer, he saw something more pertinent: a man frantically saddling a horse. The animal was fidgety with excitement from the roaring battle to the south. It kept moving while the man was tightening his latigo strap.

Tom crept closer and saw another man bridling another nervous horse. Those were the only two he could make out right then, but as he stepped over to a big bush to take aim, a third man closer to the forest shade saw him, yelled out and fired. Instantly, the other two men dropped what they were doing, hurled themselves sidewards and swung also to let go of blind shots. Tom dropped to one knee, aimed and fired. The nearest man was spun half around. He bawled a wild curse and fell in the tall grass. The other pair were now out of sight.

Hugh Tyre came up, sidling along with his carbine held sideways in both hands. A bullet came from nowhere, breaking a brittle pine limb above Tyre. The marshal whirled and fired blind in the general direction of the man, who had come close enough with his bullet to shower Tyre with bark and wood.

The fighting to the south seemed to be breaking up, to be shifting from one front to a number of individual duels. The outlaws didn't appear to be falling back upon their camp out in the clearing; instead it sounded as though they'd managed to get out into the forest also. Some of the sniping moved relentlessly up in the direction of McGrath and Marshal Tyre.

Tom crawled back where Tyre was and shook his head. Tyre's lips moved as he got down, too, but the words were lost in the thunderous crash of guns. But he and Sheriff McGrath had at least prevented three outlaws from escaping on horseback, and that had been Tom's reason for slipping around there; to stampede the horses if he could, thus settling their enemies afoot in an unfamiliar light-dark world of endless distances and many enemies.

Over in the clearing a man was shouting over and over that he was out of it; that he surrendered. No one paid any attention to him as the struggling battlers continued trying either to fight clear or prevent someone else from fighting clear.

Tom pointed over to where he'd downed the renegade in the grass and started crawling ahead in that direction. Tyre unquestioningly followed him. They got right up to the edge of the meadow this time, where they could see all around. Several horses, two of them saddled but not bridled, were high-tailing it north out of the meadow. Another horse, bridled but not saddled, was running with them the way wise old saddle animals often do when they have reins dragging; he had his head and neck turned slightly to one side to avoid stepping on the reins.

The man stopped yelling out there. He'd either been hit and had fallen in the grass, or he'd decided to lie down prudently and surrender later, when the opportunity to do so would be more propitious. Otherwise, though, the sniping was still going on, not as loudly or as regularly as before, but still amply thunderous to inspire respect in everyone.

Hugh Tyre pushed out his carbine, took careful aim down along the curve of the meadow and fired. Tom saw nothing until a man whose back had been to them suddenly sprang up and ran frantically out into the trees. Tyre had hit close enough to rattle that one completely.

Frank Moore's excited profanity rang out, calling on someone to drop his gun or be killed. That was the first complete sentence anyone had been able to distinguish since the fight had begun.

Tom said, "It's about over." He and Marshal Tyre turned back and crawled out to where it was safe to stand, then got to their feet. Two men were still savagely duelling off to the west. One would fire; then the other one, farther out, would fire back. This went on until McGrath and Tyre slipped a few dozen yards and saw other possemen converging, too. Then the man farthest away suddenly hurled his Winchester from behind a tree and yelled out that he'd had enough.

They told him to toss out his .45, too, then step forth. He obeyed. It was one of the hatless, shirtless men Tom had sighted earlier standing over the dead bull buffalo. He ambled forward, his face soiled and sweaty and twisted into an expression of monumental disgust. He didn't act the least bit afraid, just disgusted.

Jack Kirk came up with a torn and bloody left sleeve. With the exception of the dead man back behind the old deadfall tree, they all seemed to be there. Tom looked closely to make certain of this, then led them and their prisoner back toward the meadow.

They found two dead men in the forest and one with a broken arm out where Tom had dropped him in the tall grass. The others had gotten away, evidently, but Tom silenced the complaints about this by sending five men back for the horses, with instructions to get down there fast before the outlaws happened onto that spot and wiped out the man they'd left to guard the beasts.

Then they began a systematic search for more outlaws and found them—three wounded ones trying to hide. Hugh Tyre helped round up all the captives out in the meadow. He told McGrath that although they hadn't gotten Jay Adams, he was satisfied with the birds they had flushed. "Every one of them will have a bounty on his topknot. They'll be wanted somewhere."

Frank Moore was more motivated. "Sheriff," he asked, "how about you'n me an' Hugh sweatin' it out of them where Adams went?"

Tom nodded. That was precisely what he had in mind. With at least one renegade on the loose, Adams might hear what had happened and flee.

CHAPTER 17

AFTER THEIR CAPTIVES HAD ALL BEEN DRAGGED TO one spot, dead ones and wounded ones, they wore informed by an outlaw with a perforated shoulder that two men weren't accounted for. After their horses came up and the possemen were counted again, Sheriff McGrath sent out searchers, but he made them go on foot. Horsemen, under the circumstances, made excellent big broad targets, and with those two fugitives on foot, they'd kill more to get the horses of their pursuers than because they wanted to murder possemen.

That same big, curly-headed outlaw who'd volunteered this information also told them something else: "Adams? Hell, if you fellers come here to catch Jay Adams you're goin' to be disappointed. He pulled out right after breakfast this morning, right early."

Another wounded outlaw looked acidly at the talkative man and growled at him, "Jake, you better shut up."

No one paid much attention to this man; he'd been shot through the arm by McGrath. The limb was broken above the elbow and seemed to pain him considerably, but not so much he wasn't aware of what the possemen were digging for.

The big one with the tumbled mass of curly hair shrugged, then winced and cursed when the movement made his punctured shoulder burn with a stabbing pain. "What's the use?" he demanded of the other wounded man. "It's all over now."

"They ain't got Jay; he'll be back with the others and pull their stinkin' settlement down around their lousy ears. When that happens—an' he hears about your big mouth . . ." The outlaw showed a venomous expression to the other injured renegade, not bothering to finish a sentence whose conclusion was obvious to every listener anyway.

Hugh Tyre stepped up to the one named Jake. "What others?" he snapped. "What others did Adams go to get, Jake?"

The larger, younger man gazed down his nose at Tyre. He seemed singularly unimpressed. "You another sheriff or a constable, old-timer?" he asked.

Tyre turned back his vest. The outlaws all saw that familiar star in a circlet. Jake was now definitely impressed. He re-examined Hugh Tyre. "Well, Jay went up to another camp we got word of to hire three, four fellers to ride down into Puma Station, because he figured he didn't want to chance it that maybe folks might've seen some of us."

"What for?" asked Tyre. "What does he want them to do in Puma Station?"

Jake's eyes turned ironic. "Marshal, he wants them to make sure that there ex-girl friend of his is still in the lockup; then he figures to lead us all into the place and turn it upside down. With fifteen or so guns he could do it, too."

"Could *have* done it," said Frank Moore, glaring at the larger, wounded man. "Hey, Sheriff McGrath—"

"I know," exclaimed Tom, not giving Moore a chance to speak his piece. "Get the dead ones tied over horses, get the other ones trussed good; then let's get back down toward town."

They left the camp as rapidly as they could, but it still took them twenty minutes to move out. Tom took the outlaw named Jake up to the front of the column with him. Jake was talkative; he said McGrath and Tyre were wasting their time trying to guess where he was wanted by the law; that he wasn't wanted anywhere, but had simply come along because he was between cow-ranch jobs. He then laughed and said he'd keep that scar in the shoulder to show his kids some day and convince them being an outlaw was a mighty poor investment of a man's time.

When they asked him about Jay Adams, he first admitted that it had been Adams' name and fame that had swayed him to join the outlaw band. He then said, up close in camp, Jay Adams wasn't the man Jake had expected him to be.

"Why, we had plenty men to bust your jailhouse," he said candidly. "An instead of skulkin' around spyin' on the Meuller place, making a big play for that big blonde girl, I'd have gone down there and turned the place upside down for that money. Not Jay; he had to make a big affair with the girl and prove somethin' that way. All of us were gettin' bored and disillusioned even before you fellers busted out o' the trees at us."

Hugh Tyre asked if there was any chance of the two men who'd escaped back there getting over to the roadway where they might intercept Adams coming southward. Tom McGrath shook his head. Although the searchers they'd sent out back at the big meadow hadn't turned up any sign of the escapees, McGrath said they'd patrol the road when they got close enough and either catch the outlaws that way, or at least keep them from reaching Adams.

"Providin'," said a posseman, "them fellers try to reach the stage road. I tracked one of 'em until he clum into the malpais rock, an' it looked to me like he was cuttin' a big fat circle and tryin' to get around to head west again."

Frank Moore supported the intimation behind this. "If those two got a lick o' sense, they won't bother with Adams after what happened back there; they'll concentrate on savin' their own hides by gettin' out of the country."

Tom McGrath had thought that the most likely course for the fugitives back at the big meadow. But on the off-chance he might be wrong, and at least one of the outlaws would seek to warn Jay Adams, he'd decided on the other course; on having men head up-country from town, as soon as he'd gotten his prisoners locked up and the dead ones stowed out back in the Puma Station burial shed, where they'd repose until someone went out and dug the holes for them.

Jake and the other wounded men began to suffer a little before they reached town, so Tom headed for a waterhole and allowed them a brief respite. The other injured men were sullen, uncommunicative, and bleakly hostile. Even when possemen offered to help with bandages, they snarled at them. Jack Kirk, who'd gouged his upper arm and had also ripped the sleeve nearly out of his shirt when he'd dived into a thorny thicket, got the last thorn dug out and bathed his arm in the same water with Jake, the big, curly-headed captive. Jack said, "If you weren't in this thing all the way with Adams, what ever possessed you to try the outlaw life?"

Jake looked straight at Kirk. "Same as you," he retorted. "Same as most other riders, when they're between jobs, bored with towns, and with a little bow in their necks. Mister, don't sit over there foolin' with those little stickers in your arm an' tell me you've never wondered what'd it be like, 'cause if you do, you'll either be lyin' to me or just plain dumb."

Jack thought about that. "I don't like the idea of bein'

mistook for dumb," he replied gravely after a while, "so maybe I'll just let on like I always wanted to ride with Jay Adams or Jesse James—and *prove* I'm dumb like you did, partner. Darned dumb!"

They left the waterhole, heading for Puma Station, with the sun beginning to drop down off in the heat—hazed west. Tom asked Jake how far the camp was where Adams went for his spies, and about how long it should take him to get back down from there. Jake said he only knew what he'd heard the others talking about. "It'd be a day's ride up an' a day's ride back. How much time he kills at the camp is anybody's guess. But I figure he should be back in the Puma Station country come evenin' tomorrow, Sheriff."

"And," stated Hugh Tyre, "it's only going to take him a little while after that to discover that the world caved in on his hired guns at Mormon Flat."

Frank Moore nodded. "That son-of-a-gun'll high-tail it out of this territory so quick it'll make your head spin. He'll drop those new recruits of his like we've seen him do before, and just get swallowed up in the distance."

Tom McGrath said no more. He listened, though, while the others talked, venting resentment, making threats against Adams, and generalizing about what lay ahead. They were within sight of town, approaching deliberately from slightly down-country so that, as McGrath explained it, they could perhaps get into the alley behind the jailhouse without rousing up the whole cussed town, in which case they'd be nearly trampled by well-intentioned but thoroughly irritating citizens.

They made it, got in behind the jailhouse, and Tom McGrath had the dead men shoved in the shed across the alley. He then took the prisoners inside and told the possemen to stand by; he'd be out again directly, and they'd all go manhunting again.

There was some dubious speculation among the possemen out back as to whether they should remain right there and wait, or amble on up to the Angel's Roost. Of course, the saloon won out. What alleyway could compete with the bountiful enticements of a fine old manly saloon in summertime? They left a note stuck on the back of the jailhouse door where Sheriff McGrath could find them, and went cheerfully up the alleyway, leading their horses. They had enough to relate up there to be assured of at least three, perhaps even four, free drinks.

Inside the jailhouse when the pair of sulky wounded men pulled away after having their injuries cleaned and bandaged properly, Tom McGrath took each one by the shirt front, up close, and twisted his fist slowly. The outlaws felt their breath being squeezed off and stopped resisting. But neither of the two would volunteer a word, not even a negative or affirmative shake of their heads.

Tom turned them over to Jack to be locked up in the cell room, and offered Jake his makings so the curly-headed man could make himself a smoke. Jake was one of those people who didn't seem to fret nor worry. He was a straightforward man, wild, reckless, unafraid, and a typical range man. He handed back the makings so McGrath could twist up a smoke. As Tom later held the match for them both to light up from, Jack came back into the office, locked the cell-room door and tossed the key over to McGrath.

Jack said, "Those other two liked to have fainted. They never thought we could get anywhere near the camp, let alone shoot it up and salt down a couple."

"Thanks to the old bull buffalo," murmured McGrath,

motioning for the other two men to be seated. Jake looked surprised.

"Did you herd that doggoned buffalo onto us? Holy mackerel; I was drowsin' under a tree when I heard that critter let out a bellow that'd wake the dead. Next thing I knew, I sat up, and lookin' straight at me from a little ways was a darned old shaggy buffalo, coming like a railroad train across the meadow. I was too startled at first to believe it, but when the ground commenced shakin' from his heft, I believed he was real all right. That was a neat trick. I got to remember it."

"Fat lot of good remembering it will do," stated McGrath. "I stumbled onto that old devil in the canyon. He must've been holing up back in the forest somewhere. Jake, you'd never again come onto a buffalo under those circumstances again if you lived to be a hundred and ten years old."

Jake nodded. Then he switched the subject entirely. "Say, Sheriff, how about me ridin' with your posse after Adams? I'm a fair hand with a gun—anyway, this perforated shoulder's on the wrong side to slow me down anyway. I can—"

"You can go to jail," stated Tom, and scooped up the keys off his desk. "Come along."

Jake protested, but Jack Kirk gave him a shove to his feet. They took him through into the cell room, and directly in front of him, leaning on their bars, Ruth Hall and Helen Meuller saw him first. Then the other members of the smashed outlaw band also saw him. The last two men to be locked up were sitting in their cell glaring hate at Jake and his captors. One of them said to the curly-headed man, "You tell 'em everything you know, Jake. You got any idea what's goin' to happen to you?"

Jake paused out front of that cell. "Sheriff," he said, "put me in with these two."

Tom shook his head and gave the larger man a light push. He meant to lock him up with the first two renegades in the farther cell. But Jake didn't move. He was gazing steadily in at the vicious-eyed pair.

"In here," he repeated, refusing to budge even when Jack Kirk reached for the shoulder with the puncture in it. "These two've been talkin' big and tough ever since they joined up. I think someone ought to teach 'em manners. Open her up, Sheriff; this here's my cell."

Tom opened it, closed it after big Jake, and quizzically gazed at the other two. "Let's see what's goin' to happen to him," he said, and returned to the cell where Ruth and Helen were patiently waiting, consumed with curiosity. He told them of the battle. Helen put a hand to her mouth and kept it there until Tom said Jay Adams hadn't been at the camp. Then she slumped and looked enormously relieved.

But Tom didn't tell them anything of his future plans. He simply promised to have food sent to them, and walked back out into the outer office. There, he told Jack he thought they ought to go round up their possemen and start the waiting hunt for Jay Adams and his fresh recruits.

"They'll be up at Astor's saloon," remarked Jack hopefully. "There'll be free drinks up there, too."

Tom looked dour. "Come on out back," he said, and led the way. Jack didn't refuse, but he didn't look delighted, either, at the prospect of more riding with a cotton-dry gullet.

They got astride, turned and walked their animals almost up to the far intersecting roadway when McGrath remembered the food, went back and called through the back door of the café for seven portions of grub to be taken to the jailhouse. When the café man came out into the dusk to see who it was giving him this order, Tom handed the man the jailhouse keys and told him not to attempt feeding those people in their cells unless he had at least three armed men with him to see that he wasn't jumped.

The café man took the key and stood in his rear door. way gazing after McGrath and scratching his balding head. He muttered after the retreating pair of horsemen in a low tone, "If you think I'm goin' into that jailhouse alone with all them brigands you got locked up in there, mister, you're as loony as a coot."

The day was almost completely gone now. Dusk was sweeping in quietly and inexorably. The far-away curve of mountains was little more than shadow against the paler heavens. Neither Tom nor Jack had had a decent mouthful of food since early morning. If that troubled Jack—it did—Tom McGrath seemed scarcely conscious of being hungry at all.

CHAPTER 18

THE POSSEMEN WERE READY TO GO AGAIN, BUT they managed to delay Sheriff McGrath's departure long enough to rustle from several sources food which they stuffed either into their mouths, their pockets or their saddlebags. One or two also secreted ponies of rye whisky.

Hugh Tyre, getting astride a rented horse from the livery barn because he felt his private mount had been used enough lately, critically examined the ears of the fresh beast and pursed his lips. Frank Moore, also on a livery beast, didn't try to fathom the intentions of the new horse under him. He waited until all the others were together again, then stepped across leather and said to McGrath he wasn't convinced of the need for all this, since Jake had told them Adams wouldn't arrive back in the area until the following evening.

Tom's retort was pithy. "And suppose Jake's wrong. Or suppose Adams didn't lie over up there, but got his men and started back two, three hours ago?"

Moore had no more to say.

They left town, riding north. A considerable crowd of men came forth to watch them ride off. Some had asked to be taken along. Tom was satisfied with the same crew—less one—that he'd fought the outlaws with.

He took them up four miles above town and halted them. "There's a fair chance Adams won't head straight for town with these recruits of his. We'll split up here and get strung out from east to west. It's not likely he could return before sunup, if he makes it by then. All the same, keep quiet out there and keep watchful. If you see anyone, even a solitary rider, pass the word down the line. But if you see four or five armed men heading either toward Puma Station or toward that camp they had in the forest, ride for the rest of us." He paused, looking them over. "And if you hear gunfire, come on the run."

Frank Moore went poking off toward the west rangeland with the others, but Hugh Tyre remained with McGrath for a while. As he said, there wasn't any rush to get into place, and he had a couple of things on his mind. "One of 'em is: You can claim the reward, but as federal marshals Frank and I can't. What I want to know is—do you figure to claim it?"

Tom dismounted and stood beside his horse. "I hadn't

really thought too much about it, Marshal. I reckon I will, though."

Tyre regarded McGrath for a moment, then said, "And the other thing: Are you going to do anything to the Meuller girl?"

Tom shook his head, looking Tyre straight in the eye. "Why should I? What's she done, except meet an outlaw and get a scar on the soul for that?"

"Good answers," stated Tyre, and lifted his rein hand. "I'll head on west a ways and join the others, but this is likely to be a long wait."

"Maybe," conceded McGrath. "And maybe, if those men Adams went after aren't as far off as Jake seemed to think they were, it might not be so long a wait."

Tyre turned his horse and let the animal stroll west. The night thickened around them all, and although a couple of weeks earlier it would have been cold after midnight, that wasn't true of this particular night.

Tom made a cigarette, let his horse drag the reins as he grazed, and leaned upon a dusty old tree looking into the eye of the night, listening to the endless silence. The only thing he felt at all confident about was that Jay Adams would return. But he didn't put this down entirely to Adams wanting the stolen money back, because he'd figured the notorious outlaw's greater weakness—his vanity. He'd come back, and he'd do his darndest to upset the entire countryside, not just to get the money back, but also to demonstrate his power and craftiness to Helen Meuller.

Tom dropped the smoke, stamped on it and shook his head. There was no more fatal weakness in men than vanity. If he had this figured right, Adams' vanity was going to be the death of him before another sunset.

The wait was long, in one way, and short in another

way. It was just before dawn when Hugh Tyre came riding over to say he'd just heard from one of the possemen on west that Moore and two others had detected the sound of riders passing down from the north, but they weren't heading toward town at all; they were angling over toward the outlaw camp.

When Tom hesitated, thinking that if this just happened to be a party of range riders on their way somewhere in the performance of their daily routine, and he with his men all congregated off there to the west, then Adams just might come down the stage road and have a clear route into town. When Tyre asked him what the problem was, Tom caught his horse, got astride and said, "If we're wrong, we may not get a second chance. Come on; I reckon any action is better than no action."

They went west at an easy lope, picked up a man sent over to steer them in, and heard from this man that one of the riders out there coming from up north was drunk; it was this drunk man's senseless singing which had permitted the possemen to pinpoint the location of the strangers.

But by the time Tom and Hugh Tyre came up with all the others, sitting in a gloomy place with an arroyo on their left, all they could detect was the faintest sound of ridden horses passing well west of them toward the distant forest. Tom was once again assailed by doubts, but he transmitted none of these to the others. When Frank Moore edged up to suggest they make a run on the unsuspecting outlaws, Tom shook his head, turned and led them in a silent walk straight south for a mile, then turned west and headed straight for the yonder forest, boosting his animal over into a lope. This was grassy range down here; their noises would be minimal.

He hoped the separating distances would do the rest, and evidently they did, because when the trees loomed blackly uneven against the velvet heavens, McGrath halted his men and heard, far off up-country, the faint wailings of a drunken man's song.

He knew this country well. Adams would turn into the trees about at the same angle he was now riding. As long as he wasn't apprehensive, he'd make no deviation. "Leave the horses here, and bring along your carbines," he said, dismounting.

They followed through the shadows and settled night which lay thickly beyond the first fringe of forest, heading straight up toward that off-key singing. When it was loud enough to pilot them almost to the place where Adams would enter the forest, they heard someone wearily curse the singer, who then turned garrulous and bitter, after he stopped his noise, and demanded to know what kind of men didn't appreciate music.

Tom didn't see them until Jack Kirk grunted and pointed out where watery starlight showed a dark-moving blur angling toward a gap in the trees a hundred yards farther north.

Tom motioned for the others to come along and stepped warily ahead. He wanted to catch those men before they reached the dark forest.

The rendezvous was close now; McGrath's companions were painstakingly making their way toward the opening in the forest, silent as Indians, while the horsemen out there rode almost drowsily to meet them. The sun wasn't making much of a dent in the dark pre-dawn, although it was lighter beyond the forest where relatively open country lay. There was a belated chill, too, which made fingers slightly stiff and tightened the men's back muscles.

Tom halted, slashed downward with his left arm, and set an example by kneeling close to a maverick black-oak growing among all those pines and firs. He could see the oncoming horsemen very plainly now. There seemed to be five of them, although, bunched up as they were, it was difficult to tell just yet. He cocked his carbine and waited. Around him among the trees other men also eased back the hammers of guns. Jay Adams didn't have a chance.

McGrath leaned down when Hugh Tyre put his lips to the sheriff's ear. "The tall one with the pearl-grey hat—that's Jay Adams."

Tom looked longest at this one, interested, curious, and cynical. All he could see of Jay Adams' face, because the hat brim shielded it, was a heavy set of curving lips and a finely made nose that flared slightly at the nostrils. But he didn't have to see any more; he'd already been told several times the outlaw chieftain was a handsome man. He raised his Winchester, took a loose rest against the oak tree, and counted the yards separating them, then the feet. Then he dropped his cheek to the gunstock, felt the quick, sharp bite of cold steel, and said, "Stop right where you are, Adams!"

He afterwards said he hadn't thought they'd stop when he told them to. He was right; they didn't stop. Even the drunk one out there reacted to that unexpected voice in the forest fringe like a rattlesnake. They cursed and whirled downward from the saddles, firing with astonishing speed and accuracy toward the sound of McGrath's voice. Tom lost his target; Adams was on the far side of his horse, using the beast as his shield. He fired off one round over the seating leather; then the startled horse violently shied and ran straight up into the forest with the other panicked animals.

Tyre was shooting his six-gun, his Winchester abandoned on the ground. Bullets made solid sounds striking trees. Orange flame lanced outward from the trees all around. If the outlaws had thought there was only one man in there, within seconds they knew this was not so.

One outlaw tried to make a run to the rear toward some bushes. His movement seemed to catch all the attention of McGrath's possemen. They caught him ten yards short of his objective and spun him one way, then the other way, with their lead. He dropped.

Another outlaw tried a different flight; this one fired as fast as he could, shooting at muzzle blasts, and ran zigzagging it straight for the forest. He almost made it. Hugh Tyre shot a leg out from under him, which wouldn't have been fatal, but as he staggered almost to a halt, bullets ripped into him with a meaty sound. He too fell and lay still.

Adams had lost his hat. Tom McGrath saw his face more clearly now. The notorious outlaw looked frightened; at least that thought came to Tom as he fired, scuffed up dust in front of Adams and corrected his aim to fire again. Adams rolled, realizing he was under personal attack. Tom anticipated the roll to the left and caught Adams hard with his next shot.

There was another man directly behind Adams who saw the outlaw leader wilt. This man yelled for quarter and flung away his gun. Off to the northwest another man, the only one still firing, was backing up as he fought, wiggling farther and farther away. He was the only one still willing to make a fight of it.

Tom raised an arm. "Hold off," he yelled to the possemen. Their firing ceased. Tom dropped his arm, watched the pair of uninjured men a moment, then addressed the one who'd surrendered.

"Get up, mister, shove your arms high, and walk over toward the sound of my voice."

The outlaw clambered dustily and stiffly to his feet to obey. Tom tapped Hugh Tyre and pointed. Hugh nodded; he was to watch this one. Tom then called to the remaining one out there to the northwest. "You're it, cowboy," he said. "You out there trying to wiggle away. It's up to you; either drop the gun and stand up or get wiped out. Make up your mind fast."

For a long moment there was no response. Possemen cocked guns and raised them again. The outlaw then cried out. "All right, ya bunch of bushwhackers. Here's m' weapon." He flung something heavy as hard as he could toward the forest. He then stood up, not too steadily, and began cursing in a singsong voice as he lurched ahead. This was the drunk one, apparently, and he kept up his verbal abuse until three possemen rose up out of the forest fringe and reached to haul him into the darkness and dump him, hard, on the ground. That seemed to have a sobering effect; he stopped swearing and lay there looking up into the barrels of the cocked pistols.

Tom got to his feet, set the carbine aside and walked out with Jack Kirk and Hugh Tyre to where Jay Adams lay. They rolled the famous outlaw over onto his back, expecting him to be dead. He wasn't, but he was dying. McGrath's bullet had caught him anglingly at the point of the shoulder and had carved its way through his lungs, and lower.

He seemed to recognize Hugh Tyre, but he looked longest at Tom McGrath. His lips moved slightly. Tom dropped to one knee, listening.

"Lousy—lawmen . . ." Adams whispered in a fading

voice. "Dirty lousy. . ."

Tom bent for a closer inspection of the wound. As he straightened back he looked up at Hugh Tyre. The federal officer was waiting for Adams to die without a particle of compassion in his smoky eyes nor across his wintry face. "If you want to ask him anything," said Tom, "you'd better hurry."

"Nothing," murmured Tyre. "There's nothing I want to know that he'd tell me. Let him die."

He did die, still with curses on his handsome lips, looking from dark eyes with a hating spirit at Tyre and Tom McGrath.

Jack took some of the men and went after the horses. It took them almost an hour to round up the loose horses of the renegades. By then the drunken man was stone-sober. He was bitter, too. "Adams," he said, spitting out the name. "He was so smart. He knew everything: how to get his cache back, how to get his revenge. To hear him tell it when he come to our camp, he had the world by the tail on a downhill pull. An' look at him—the lousy clown with his clean shirt and pearl-grey hat. *Look* at him!"

They tied the dead ones belly-down, tied the captives astride, then mounted up to head on back. The sun was making its first faint paleness over against the eastward sky. No one had much to say on the ride back, but Hugh Tyre asked a question just before they reached town.

"How about the husky little girl?" he asked. "What you figure she'll do, Sheriff?"

Tom considered the soft-lighted rim of the world as he answered. "I don't think she's got anything in Idaho to take her back. I'm hopin' she'll stay."

"Stay?" said Tyre. "Stay in Puma Station? What is there for her in a place like Puma Station?"

"Well, I'm not just sure," drawled Tom McGrath, looking now at the town. "But I reckon she could get married. Lots of men around who're sick of restaurant food an' livin' in rooming-houses, Marshal. Lots of men around Puma Station who'd admire mightily having someone like that to look at, come evening, and go on picnics with."

Hugh Tyre bent a little in his saddle to peer closer. "I'll be damned," he gasped, running the words all together. "I'll be *double* damned. I reckon you're right, Sheriff, and I sure hope you're successful in talkin' her into it."

They entered town from the west, as before, avoiding contact with the few folks abroad at sunup. Also, as before, they deposited the dead ones, including Jay Adams, in the shed out back. Then the possemen broke up, heading for home and sleep. Even Tyre and Moore left Tom McGrath and Jack Kirk out there in the alley.

Jack said, gazing at the forbidding rear door of the jailhouse, "Who's going to tell Helen we had to kill him?"

Tom fished for his key, saying, "I'll do that." He hesitated, squaring his shoulders, then opened the door and motioned the sullen prisoners in first. "You go get her paw and tell him to come with his buggy and take her home."

Jack walked away. Tom went inside, told his prisoners to sit down, and turned up the lamp. He then made a smoke and stood a moment just gazing at the cell-room door. He'd never liked that door, but tonight—or rather this morning—he liked it less than ever.